High-performance Buildings:

A Guide for
Owners & Managers

High-performance Buildings:

A Guide for
Owners & Managers

Tony Robinson, MS

LONDON AND NEW YORK

Published 2020 by River Publishers
River Publishers
Alsbjergvej 10, 9260 Gistrup, Denmark
www.riverpublishers.com

Distributed exclusively by Routledge
4 Park Square, Milton Park, Abingdon, Oxon OX14 4RN
605 Third Avenue, New York, NY 10017, USA

First issued in paperback 2023

Library of Congress Cataloging-in-Publication Data

Robinson, Tony, MS.
High-performance buildings: a guide for owners & managers / Tony Robinson, MS

   pages cm
  Includes bibliographical references and index.
  ISBN-10: 0-88173-646-5 (alk. paper)
  ISBN-13: 978-8-7702-2290-7 (electronic)
  ISBN-13: 978-1-4398-5199-9 (Taylor & Francis distribution : alk. paper) 1.
Sustainable buildings--Design and construction. 2. Buildings--Performance. 3.
Buildings--Energy conservation. I. Title.

  TH880.R65 2013
  658.2--dc23

2013032999

High-performance Buildings: A Guide for Owners & Managers / by Tony Robinson
First published by Fairmont Press in 2014.

Routledge is an imprint of the Taylor & Francis Group, an informa business

Publisher's Note
The publisher has gone to great lengths to ensure the quality of this reprint but points out that some imperfections in the original copies may be apparent.

ISBN 13: 978-87-7022-915-9 (pbk)
ISBN-13: 978-1-4398-5199-9 (hbk)
ISBN- 13: 978-8-7702-2290-7 (online)
ISBN- 13: 978-1-0031-5159-3 (ebook master)

While every effort is made to provide dependable information, the publisher, authors, and editors cannot be held responsible for any errors or omissions.

Table of Contents

Preface

Buildings, their occupants, and the environments they operate in are complex systems. The sphere of analysis required to evaluate their integrated financial, energetic[1], and environmental performance has not yet been fully determined and the pace of change in building technology, the regulatory arena and energy modeling is rapid. The choice of system boundaries determines the outcome[2] and we are still learning how to define those boundaries. Accordingly, any work on the subject of high-performance buildings cannot promise to be comprehensive or entirely up-to-date. A book that addressed all of the issues in a comprehensive manner would run to more than a thousand pages, so inevitably some probably significant technologies must be left out. Accordingly, this is primarily a book of principles, a survey of major ideas and a template to be used for guidance on what could be various paths to the goals of high-performance. In a certain sense, then, it is meant to be a comprehensive introduction to what is a very large subject. The interested reader can do further research in particular areas as the situation requires. Although this book has a major focus on existing and commercial buildings, it can also be used by the designer or building engineer of new and/or residential buildings.

Our approach will be program and product agnostic. In other words, we will treat the subject from the standpoint of a scientific observer, examining the most important evidence we have at this time, with our goal to render the clearest and most objective exposition of the methods and materials available for the purpose. It is important to state at the outset, that just as there is no one energy source which can provide for all of our needs, there is no one method of analysis which can provide the solutions for making all buildings high-performance. Additionally, the owners and managers of buildings have their own specific goals which may not dovetail with particular engineering programs, ratings or reporting systems. Nevertheless, they must remain well-informed on the evolution of building technologies, energy codes and environmental regulations. Only in this way can they make good decisions about how to prepare their buildings for compliance and financial performance well into the future.

May, 2013, Dallas, TX

1. *Energetics*: the branch of physics which studies energy transformations
2. *Energy in Nature and Society*, Smil, V., MIT Press, 2008, p. 274

Acknowledgements

There are many people I owe a debt of gratitude to over the past 25 years in my career who have helped to make this book possible. I've had some great teachers in the field, in the office and in the classroom. I would like initially to thank my colleague Dr. Eric Woodruff, President of Profitable Green Solutions and former president of the Association of Energy Engineers who is a real warrior for the cause of high-performance and energy engineering. Eric referred me to Al Thumann at AEE and the publisher CRC/Fairmont Press. A big thanks as well to all of my co-authors: Dr. Victoria Chen of the University of Texas at Arlington; Jyl deHaven, MS, of Green Collar Vets and EARTH-NT; Tim Janos, C.E.M., of the Association of Energy Engineers; Dr. Pin Kung of Cheng Kung University, Taiwan; Jason Thompson, P.E., of Burton Energy Group; and Dr. Paul Tinari, P.E., of the Pacific Institute for Advanced Study.

I would also like to thank my former graduate adviser and teacher, Dr. Michael Kozak (recently retired), for many years the chair of the Department of Engineering Technology at the University of North Texas where I did my master's work. Mike was a great teacher and made sure we knew how to write a technical paper. Other than requiring us to use the APA format and make no mistakes, his direction was simple: "Tell me what you're going to say. Say it. Tell me what you just said."

Over the years I've done plenty of research into building and materials science, manufacturing engineering, construction processes and energy analysis, but a lot of my knowledge has also come from being in the field and seeing what works and what doesn't. I've been fortunate to have involvement in the construction industry at many different levels: field personnel, subcontracting, general contracting, design, engineering, manufacturing, distribution, construction management, manufacturer's representation, product design and development, on-site testing, and consulting.

I've also had the opportunity to work in a lot of different environments: Maine, Boston, New York City, California, Arizona and Texas. I've had some really great mentors in the industry too, people I started out with in business who've fought the good fight for high-performance (even before it was called that) for a long time: Brent Olson, Steve Charlson and Chip Marvin. There are also friends and fellow writers who've taken the time to read what I write and give me good advice: Richard Wirick, Dr. Kenneth Dekleva and Dean Ferguson.

xi

Author & Editor's Note

A book with this kind of scope requires research over many years into the works of hundreds of organizations and individuals. Since I have been in the industry for a long time, a lot of the knowledge expressed here has been accumulated over many years of experience. Although I have consulted numerous printed works (many of which I own) that contain the exposition of enduring principles, the vast majority of research has been conducted over the Internet. Some of this is obviously out of convenience, but it is also because the most up-to-date information on technologies, methods, materials and regulations is to be found on the web. With a subject as big and broad as high-performance buildings, the web is a two-edged sword. There is a vast quantity of information available, and the task is not to be comprehensive, but to cull through hundreds and hundreds of sources to find the clearest, the most well-reasoned and the most authoritative works. So I have footnoted almost all of the works cited with web addresses, so that e-versions or PDF versions of this book will allow instant access to the original sources and promote further research by the reader. Full titles of the organizations or printed materials, if not in the footnotes, can be found in the Sources section at the end of the book. A few of the works listed in the Sources have not been specifically referred to in the text but are closely related to the subject matter and offer opportunities for further research.

If I have paraphrased an author's work or reproduced a segment of text, I have made every attempt to trace it back to the original source, cited it in the usual way, or put the text in italics or quotation marks. In order to explain clearly the concepts and principles involved in high-performance buildings it has been necessary to use charts, graphs, tables and images from the internet in the text of this book. I have made every effort to include those items when they are in the public domain wherever possible and include the reference. Inevitably some of the items may be subject to copyright, and I have tried to employ them in the spirit of the doctrine of fair use. In other words, my intention has been to use them sparingly to educate and illustrate the ideas, and to inform and to encourage the reader to pursue the author's original work further and to collaborate on what is a long-term and very important endeavor.

In a number of cases I have referred to specific manufacturers or products to illustrate a particularly good example of something. This does not constitute a recommendation or an endorsement by me, as I do not have any relationships with any of the manufacturers named in this book.

Contributors

Tony Robinson, MS, Editor, has been involved with energy efficiency and sustainability in the built environment for more than 25 years, including broad experience in product design and development, engineering technology, manufacturers' representation, contracting, and building science. Among other pursuits, he has worked on the development of sustainable landscape construction, modular greenhouses with drip-feed irrigation, daylighting, passive solar design, technical analysis for architectural products and the building envelope, and systems engineering for decision frameworks in "green" building. He is president of Axis Building Consultants and adjunct professor in Environmental Sustainability at Southern Methodist University where he teaches two courses: *Energy & Economy: The Sustainability Factor, and Environmental Sustainability: Current Issues in Energy, Politics and Economic Development.*

His writings on energy and buildings have appeared in publications of The Industrial and Systems Engineering Research Conference, The Clean Technology & Sustainable Industries Organization, The IEEE—International Symposium on Sustainable Systems & Technology, The National Career Development Association, The Association of Energy Engineers, and The American Solar Energy Society. Tony has presented at technical conferences around the US, for career counseling and architectural groups, and for universities in the North Texas area. He is a member of the Clean Technology and Sustainable Industries Organization, the Association of Energy Engineers, and is past president of the Dallas-Fort Worth California Alumni Association. He holds a BA in philosophy with a minor in English from the University of California at Berkeley, and an MS in Design, Engineering Technology and Business Administration from the University of North Texas.

Dr. Victoria Chen is Professor of Industrial & Manufacturing Systems Engineering (IMSE) at The University of Texas at Arlington. She holds a B.S. in mathematical sciences from The Johns Hopkins University, and an M.S. and Ph.D. in operations research and industrial engineering from Cornell University. From 1993-2001, she was on the Industrial and Systems Engineering faculty at the Georgia Institute of

Technology. From 2001-02, she was a visiting professor in the College of Business at The University of Texas at Arlington. She joined IMSE as associate professor in 2002 and she visited with the University of Genoa, Italy in the summers of 2001 and 2003. Dr. Chen co-founded the Center on Stochastic Modeling, Optimization, & Statistics (COSMOS) in 2008 and served as COSMOS director until 2012. She has been serving as interim department chair since 2012.

Dr. Chen has been actively involved with the Institute for Operations Research and Management Science (INFORMS), the American Statistical Association (ASA), and the Institute of Mathematical Statistics (IMS). She served as vice-president of the Atlanta Chapter of the ASA from 1995-2001, and she was program chair for the Fifth IMS North American New Researchers Conference in 2001. In 2004, she co-founded the INFORMS Section on Data Mining, and in 2005, she founded the Workshop on Artificial Intelligence and Data Mining. She has served as secretary for the INFORMS Forum for Women in OR/MS, chair of the Informs Section on Data Mining, and cluster chair for both Data Mining and Computing. She has been the Fora representative for the INFORMS Subdivision Council and the INFORMS Chapters/Fora Committee since 2012. Dr. Chen is also a guest editor for the *Annals of Operations Research* and associate editor for the new journal *Elementa: Science of the Anthropocene*.

Dr. Chen's research utilizes statistical perspectives to create new methodologies for operations research problems appearing in engineering and science. She has expertise in the design of experiments, statistical modeling, and data mining, particularly for computer experiments and stochastic optimization. She has studied applications in inventory forecasting, airline optimization, water reservoir networks, wastewater treatment, air quality, energy, nurse planning, and pain management. Through her statistics-based approach, she has developed computationally tractable methods for stochastic dynamic programming, stochastic programming, yield management, environmental decision-making, and simulation.

Jyl DeHaven, MS is the co-founder and CEO of the Energy and Resource Technology Hub of North Texas (EARTH NT). For the past 15 years, Ms. DeHaven has focused on the rapidly growing "green" industry, and the integrative approach to sustainable business models, resource planning, and corporate social responsibility.

With broad expertise in commercial real estate, she has been a primary creator and driver of the vision for sustainable real estate development in the North Texas area, and has handled fund raising on several urban infill redevelopment projects (pre-development under $10 million). Ms. DeHaven is an active contributor to the Urban Land Institute, the North Texas Council of Governments, Vision North Texas, the US Green Building Council, Commercial Real Estate Women (CREW) (past president) and Habitat for Humanity. She has served on the boards of the Greater Fort Worth Real Estate Council, Urban Forestry—co chair, the City of Fort Worth Mayor's Sustainable Building Council—co-chair, Fort Worth Professional Women—past president, the Chamber's Women Influencing Business, and the DOL Sustainable Jobs focus group.

Ms. Dehaven is an associate clinical professor at UT Arlington in the School of Urban and Public Affairs and is senior advisor to Tarrant County Community College. She is also board president and co-founder of Green Collar Vets—a non-profit organization that assists veterans returning from Iraq and Afghanistan looking for jobs in "green" industries. She holds a bachelor's degree in fine art and design from The University of Texas at Arlington and a Master of Science in Nutrition and Health from Texas Women's University.

Timothy B. Janos, C.E.M., C.E.A., C.D.S.M., C.S.D.P., B.E.P., is a life member of the Association of Energy Engineers (AEE). He served as international president of AEE in 2006 and is currently director of special projects and board chairman of the Certified Energy Manager Program. Tim was elected to AEE's Energy Manager Hall of Fame in 2007, and he is also an international instructor and trainer for AEE. He was selected to teach the first Energy Manager Certification Course for AEE in mainland China.

As principal of Spectrum Energy Concepts, Inc., Tim provides domestic and international energy consulting services for a broad spectrum of clients in industry and government: Goodyear Tire and Rubber, General Motors, Cargill, Duke Electric, First Energy, PSE&G, Johnson Controls, Honeywell, Schneider Electric, Siemens, the Cleveland Public Library and the Cuyahoga County Planning Commission. He has more than 35 years of industry experience in the application of energy efficiency to buildings, including design-build mechanical systems contracting, temperature control contracting and energy management consulting. Tim holds a BA from John Carroll University and a JD from

Cleveland State. He can be reached at tjanos@aeecenter.org or at tim@ spectrumenergy.net.

Dr. Pin Kung received a bachelor's degree in statistics with a minor in industrial management science from National Cheng Kung University in Taiwan. He received his master's degree in industrial engineering and management from Yuan Ze University in Taiwan. In 2012, he received a Ph.D. in industrial and manufacturing systems engineering from the University of Texas at Arlington. Some of his doctoral studies there focused on stochastic modeling and optimization for decision frameworks in high-performance building using building energy analysis software.

He has lectured on quality management at Cheng Shiu University, Taiwan, and is presently working on a postdoctoral fellowship in statistics at Cheng Kung University. He is serving as a green building consultant for Tien-Fa Kung Architects in Taiwan. His research interests include applied statistics, quality management, green building, green supply chain and advanced process control.

Jason Thompson, P.E., has over 18 years of experience in the energy services industry and has directed energy-related projects and programs for commercial, industrial, and institutional facilities throughout the United States and other international locations. Among other pursuits, he has extensive experience performing energy consulting services to optimize the performance of buildings through retro-commissioning (RCx) services, technical energy audits, and engineering studies. His work experience includes positions with Servidyne Systems, LLC and Atlanta Gas & Light Company.

He is currently manager for demand-side programs with Burton Energy Group, where he is also responsible for managing energy conservation programs. He has a Bachelor of Science in Mechanical Engineering with High Honors from Georgia Tech, and is a registered Professional Engineer in three states. Jason has presented at technical conferences around the US, including the World Energy Engineering Congress, federal environmental symposiums, and Energy Efficiency Expo 2010.

Dr. Paul D. Tinari, M.Ed., P.Eng., is the founder and current director of the Pacific Institute for Advanced Study, North America's first Virtual Research and Development Institute. Dr. Tinari was born in New Haven, CT, and he obtained his early education in both French and English at several international schools before completing a B.Sc. in engineering physics at Queen's University, Ontario, Canada. While still an undergraduate, he won a number of engineering design awards for his innovative work in sustainable and zero-net-energy buildings and in alternative energy technology. After graduation, he worked as a consultant to the governments of both Ecuador and the United Kingdom in the areas of energy efficient design and solar engineering.

After receiving an M.Sc. in solar engineering, he completed a Ph.D. in environmental and applied fluid mechanics at the von Karman Institute in Brussels, Belgium. His research work was sponsored by the U.S. Air Force and by NASA and focused on the design and development of an advanced concept heat transport system for the International Space Station and for the first generation of hypersonic aircraft. Dr. Tinari currently holds certifications in mechanical, chemical and environmental engineering, education and also in theoretical physics.

Chapter 1

Inputs

When I began in the construction industry many years ago, terms such as "green," "sustainable," and "high-performance" had no current usage. In general, buildings were not considered to be integrated systems from the standpoint of energy, and their different parts were often designed by specialists who usually did not talk to one another at the same time during the design and engineering process.[1] When people were concerned with building performance, they were occasionally talking about energy efficiency, but more often than not, they were talking about return on investment.

Some "forward-thinking" designers, architects, engineers and self-taught builders were actually looking back at construction techniques from hundreds or even thousands of years ago which had been abandoned. Their goal was to reconsider some of the principles of building which were used prior to the introduction of electric lighting, mechanical refrigeration and central heating, and integrate them with contemporary tools and methods. Since our pre-industrial predecessors were working primarily with local materials, natural energy flows (solar, wind & water) and human and animal power, they devised a multitude of effective techniques for constructing buildings, many of which had been discarded by the 1920s. Modern industrial societies had become based on accelerated population growth and the exponential increase in *resource flows (RF)*, i.e. the availability and the consumption of fossil fuel energy, water, and raw materials.

ENERGY

There can be no doubt that energy transformations resulting from the extraction and combustion of fossil fuels have catapulted much of human civilization to an extraordinary level of technological proficiency and production. The fact is that we will need fossil fuel energy for some time to come. There can also be no doubt that this extraction-combustion process cannot continue indefinitely to provide for all of our needs, as these resources become scarcer, more expensive and increasingly risky to obtain.

In engineering, waste is the enemy of efficiency. Consider the following "spaghetti chart" (Figure 1-1), published annually by the US Department of Energy (DOE) in association with the Lawrence Livermore National Laboratory (LLNL) and the Energy Information Administration (EIA). The quad is a quadrillion British thermal units (Btu), a unit of energy used to universalize the source measurements to allow for easy comparisons. Notice that the total amount of US energy use is approximately 100 quads, which means that we can consider the actual numbers as roughly percentages. One can see that at this point what we call renewable energy sources (solar, hydro, wind, geothermal, biomass), although growing, still make up a very small percentage of our total energy use (less than 10%). One can also see at the top right of the chart: *Rejected Energy 55.6. What this means is that more than 50% of the energy we produce from all sources is lost from low-efficiencies or waste* and of that, 26.6 quads, or 48% of that rejected energy is attributable to energy used in electricity generation for residential, commercial and industrial use. Clearly, there is much room for improvement just in the category of waste reduction in energy use alone.

As recently as 2008-9, The US was the primary energy consumer in the world. In 2010, China took the lead with the US consuming approximately 97.8 quads of primary energy, about 19% of total global demand (Figure 1-2). Approximately 40% of U.S. primary energy was consumed in the building sector, 44% more than the transportation sector and 36% more than the industrial sector. Of the roughly 40 quads consumed in the building sector, homes accounted for 54% and commercial buildings accounted for 46%.[2] The building industry, extending from material acquisition, processing, manufacturing, transportation, construction and disposal, is one of the most energy-intensive industries in the world and is therefore ripe for widespread energy efficiency and high-performance measures.

With the US population at approximately 300 million and the global population at approximately 7 billion, this means that about 4% of the world's population uses about 20% of the energy. We can also look at per capita energy consumption in Figure 1-3.

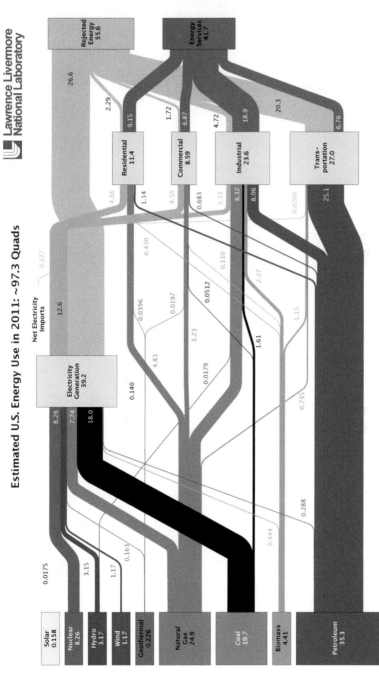

Estimated U.S. Energy Use in 2011: ~97.3 Quads

Lawrence Livermore National Laboratory

Figure 1-1

Source: LLNL 2012. Data is based on DOE/EIA-0384(2011), October, 2012. If this information or a reproduction of it is used, credit must be given to the Lawrence Livermore National Laboratory and the Department of Energy, under whose auspices the work was performed. Distributed electricity represents only retail electricity sales and does not include self-generation. EIA reports flows for non-thermal resources (i.e., hydro, wind and solar) in BTU-equivalent values by assuming a typical fossil fuel plant "heat rate." The efficiency of electricity production is calculated as the total retail electricity delivered divided by the primary energy input into electricity generation. End use efficiency is estimated as 80% for the residential, commercial and industrial sectors, and as 25% for the transportation sector. Totals may not equal sum of components due to independent rounding. LLNL-MI-410527

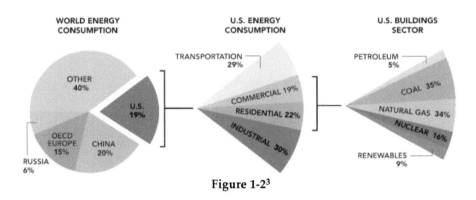

Figure 1-2³

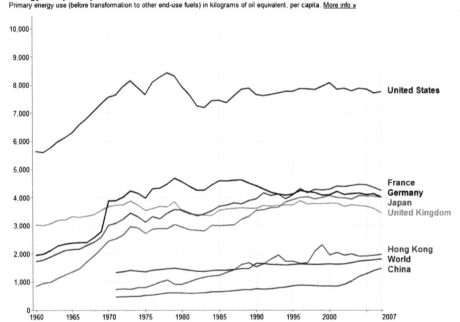

Figure 1-3

Although usage has recently leveled off, the US uses approximately twice as much energy per capita as Western Europe/OECD. Some of this has to do with geographic area, population distribution and the greater distances people have to travel across the country, but much of it is simply behavioral.

Another critical factor commanding our attention is that outside of water, the most important commodity to the functioning of a modern industrial society is fossil fuel energy, and of those energy sources, oil is the most essential. In his monumental history of the oil industry, *The Prize*, Daniel Yergin[4] demonstrates conclusively that the price and the control of oil played a major role in dictating the course of events in the 20th century. Consider the graph from WTRD Economics in Figure 1-4.

From just after WWII to 1973 and the Oil Embargo, the price of oil travelled in a fairly narrow range and was relatively stable. There are numerous reasons for that which are beyond the scope of the discussion in this book, but suffice it to say that those days are almost certainly over. With a very volatile price of oil, regardless of where we get it, planning for energy use or business costs is extremely difficult. This simply makes it all the more important to cut waste and reduce consumption, increase efficiency, and diversify energy resources as soon as is practically possible. It is just as risky to be overly dependent on one particular energy source as it is to have to acquire it from a politi-

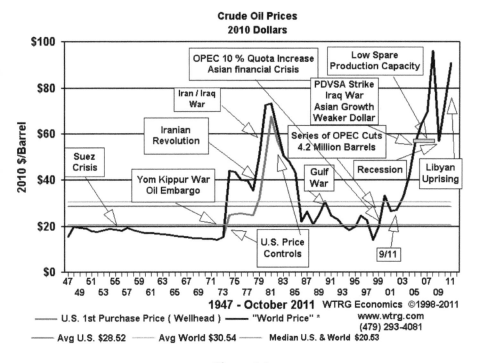

Figure 1-4

cally unstable region or from geographical locations subject to major weather disruptions.

WATER

The only commodity more important than energy to the functioning of human society and the buildings we build is water. Water is surely a commodity, as approximately 90% of consumption in the US for buildings now comes from public supplies (water utilities). The USGS and the DOE regularly publish data on water supplies and consumption, and the message is clear: the building sector and the energy infrastructure required to run it use a large percentage of the available water, and not very efficiently either (we don't even know what percentage of water use is wasted from leakage, for example).

Calculating the total water usage of buildings and the construction industry is complex and has not yet been completely determined.[5] The pie-chart in Figure 1-5 from the USGS does not specify buildings as a category, but it gives us a good overall picture, and we can do a rough extrapolation. Public supply is almost entirely dedicated to buildings. Approximately 70% of thermoelectric power (electrical energy generated from the combustion of fossil fuels and uranium) is used for buildings. This means that at least 40% of total water usage is dedicated to the operation of buildings (public supply + thermoelectric generation), although we do not know what percentage of water is used in the manufacturing of building materials and the processing of C&D (construction & demolition) wastes, which would ultimately add to the total.

Another way to look at this issue (similar to the spaghetti chart for energy in Figure 1-1, except with quantities in trillions instead of quads), is from some very important research done recently at the University of Texas at Austin (Figure 1-6).[7]

This indicates that about 12.5 quads of US energy is devoted to the residential, commercial and industrial sectors for water services as shown, with approximately seven of those quads in rejected energy. This is a very significant percentage and demonstrates major opportunities for energy and waste reduction in water usage in the building sector.

The US has a more highly developed industrial society than most other nations and this, in part, accounts for higher aggregate usage in energy and water. As with energy, the US also uses more water per

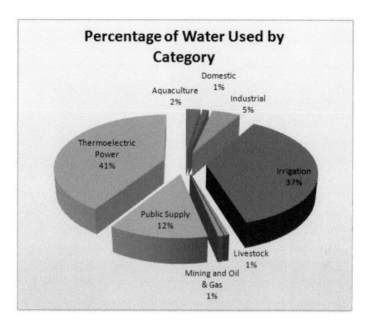

Figure 1-5[6]

capita than any other nation, as evidenced by the graph from the UN Human Development Report (Figure 1-7).[8] We operate a lot more machines, vehicles and technologies than other nations, but we can still do a lot better job handling supply and demand efficiencies.

MATERIALS

The construction and building sectors are the biggest consumers of raw materials of any industry in the US. On this subject, The Center for Sustainable Systems at the University of Michigan has published a very useful summary of US material use.[10] Some of the conclusions from the fact sheet:

_ From 1996 to 2006, U.S. raw material use increased by 29%.
_ Construction materials, including stone, gravel and sand comprise the largest component, around three quarters of raw materials use.
_ The use of nonrenewable materials has increased dramatically (from 59% to 95% of total materials by weight) over the last century as the U.S. economy shifted from an agricultural to industrial base.

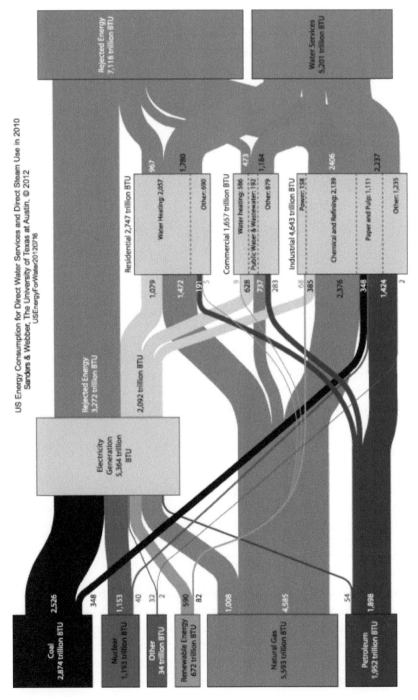

Figure 1-6

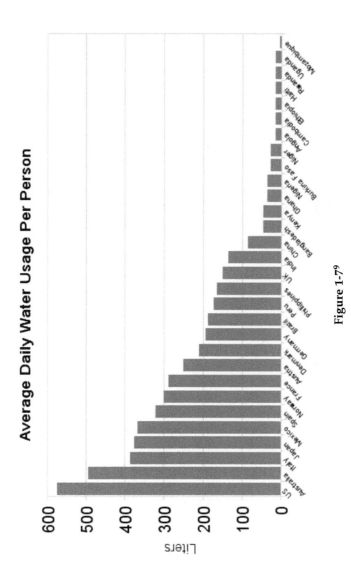

Figure 1-7[9]

Figure 1-8 shows the tremendous increase in materials consumption dedicated to construction (highest peaks in light gray) at over 3500 million metric tons, with the next highest category (in dark grey) industrial minerals at just under 1000 million metric tons.

The inevitable question which follows from the evidence of total construction materials consumption is: How much of it is wasted?

"The U.S. generated approximately 254 million tons of municipal solid waste (MSW) in 2007[12]... Building-related construction and demolition (C&D) debris totals approximately 160 million tons per year, accounting for nearly 26 percent of total non-industrial waste generation in the U.S. Combining C&D with MSW yields an estimate that *building construction, renovation, use and demolition together constitute about two-thirds of all non-industrial solid waste generation in the US*. Sources of the building-related C&D debris waste stream include demolition (accounting for approximately 48 percent of the waste stream per year), renovation (44 percent), and new construction (8 percent)."[13]

Although serious efforts have been initiated in the past ten years to streamline manufacturing processes, reduce the consumption of non-renewable materials and cut the waste stream of C&D materials by recycling/reprocessing, more than 70% of the waste still goes to the landfills.

CONCLUSIONS

Because the building industry is clearly one of the most energy, water and material intensive industries in the world, and is responsible for the majority of resource flows, inevitably the question of environmental impact arises. It is worth stating the obvious: every time we become more efficient and use less of a material or cut down on a process that has a negative impact on the environment from the standpoint of pollution, finite resource depletion or greenhouse gas emissions, then *we are improving the "environmental bottom line" at the same time that we are improving the financial bottom line*. Consequently, it can be argued at this juncture that *waste reduction is job #1*, so we in the building industry have a major part to play in contributing to the optimization of energy and water use and the elimination of waste here in the United States. All of this just makes the case for high performance in buildings all the more compelling.

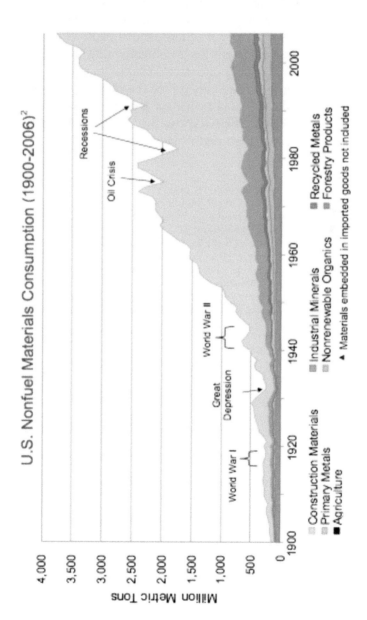

U.S. Nonfuel Materials Consumption (1900-2006)[2]

Million Metric Tons

Construction Materials
Primary Metals
Agriculture

Industrial Minerals
Nonrenewable Organics

Recycled Metals
Forestry Products

▲ Materials embedded in imported goods not included

World War I

Great Depression

World War II

Oil Crisis

Recessions

Figure 1-8[11]

The overall intention then with this book is to provide the building professional, the building owner or manager, the student, or even the committed layperson, a map for navigating what can be a very complicated process. It is primarily for those who already own or manage buildings but could also benefit those contemplating the construction of a new building. Where does one start? What is "high performance?" How do we consider the entire life-cycle of the products and processes of building? As we shall see, only when all aspects of a building's function and performance are reviewed can an "integrated systems" approach be applied and a "high-performance" result obtained.

References
1. Unfortunately, this is still too often the case.
2. Buildings Energy Data Book: http://buildingsdatabook.eren.doe.gov/
3. http://buildingsdatabook.eren.doe.gov/ChapterIntro1.aspx
4. The Prize, Daniel Yergin, Free Press, 1992.
5. What appears to be conspicuously absent from the literature is a Total Consumption figure for water usage by all sectors of the construction industry and existing buildings. This would include a summation of all water used in raw materials extraction, processing and building materials and equipment manufacturing, all water used for any energy generated to operate buildings, all water used in the operation of buildings, and all water used for the processing of construction and demolition (C&D) wastes.
6. http://water.usgs.gov/watuse/
7. *Evaluating the energy consumed for water use in the United States*, Kelly T Sanders and Michael E Webber, 2012, Department of Mechanical Engineering, The University of Texas at Austin, 204 E. Dean Keeton St. Stop C2200, Austin, TX 78712-1591, USA, available at:
   http://iopscience.iop.org/1748-9326/7/3/034034/pdf/1748-9326_7_3_034034.pdf
8. http://hdr.undp.org/hdr2006/pdfs/report/HDR06-complete.pdf
9. Consider this statistic (which demonstrates a huge opportunity for innovation in irrigation equipment and operation): "Of the 26 billion gallons of water consumed daily in the United States, approximately 7.8 billion gallons, or 30 percent, is devoted to outdoor uses. The majority of this is used for landscaping. The typical suburban lawn consumes 10,000 gallons of water above and beyond rainwater each year. "http://www.epa.gov/greenbuilding/pubs/gbstats.pdf
10. http://css.snre.umich.edu/css_doc/CSS05-18.pdf
11. Source: http://css.snre.umich.edu/css_doc/CSS05-18.pdf
12. US EPA, Municipal Solid Waste in the United States. 2007 Fact and Figures. http://www.epa.gov/osw/nonhaz/municipal/pubs/msw07-rpt.pdf
13. Municipal Solid Waste in the United States: 2007 Facts and Figures. Office of Solid Waste, U.S. Environmental Protection Agency. October 2003. http://www.epa.gov/epawaste/nonhaz/municipal/msw99.html

Chapter 2

Definitions

In the last 10-15 years as the building industry has become more concerned with energy performance, numerous terms have entered common usage which are used on a regular basis but whose meanings are not entirely clear: "green," "sustainable," "energy efficient," and more recently, "high-performance." Some of the uses of these terms are only vaguely scientific, or at the very least leave a lot of room for interpretation. What is a *green* building code? What constitutes a *sustainable* process? Why is one process more *energy efficient* than another, especially if the two processes use different sources of energy? Why is one building *high-performance* and another is not? High-performance compared to what: low-performance? Average performance? How does the term high-performance relate to the other three? These are all obviously to some extent relative terms and they depend for their meanings on baselines for comparison.

It is important to define our terms as carefully as we can before we proceed to examine prescriptions for building techniques or offer assessments of various paths for achieving our goals based on these terms. Since our approach is primarily concerned with building science and engineering, we need good working definitions. The point is not to pass judgment on the efficacy of the terms and how they are currently used, it is to understand clearly what we mean by them in this book.

Green is a general term which has been used for many years to describe an approach to building which emphasizes the mitigation of environmental impact through waste reduction, energy efficiency and the reduction of greenhouse gas emissions. It is to some extent a popular term as well which often contributes to its broad and flexible meaning.

13

As a result, there is no universal working definition of the term, and certainly no universal way as of yet to quantify what is *green* and what is not. This does not mean that the term is not useful, and in fact as we shall see it is used by code authorities and rating agencies in their documents. So we have for our purposes:

> *Green means any approach to building which mitigates*
> *environmental impact through waste reduction and energy efficiency.*

SUSTAINABLE

Sustainable is another term which seems to have a fairly wide variety of interpretations. The *Bruntland Report* published in 1987 defines it thus:

> *Sustainable development is development that meets*
> *the needs of the present without compromising the ability*
> *of future generations to meet their own needs.*[1]

This is obviously much like a statement of policy, an ethical statement which urges us to conduct our behavior in a way now that is considerate of others in the future. As such, it is a very important goal. But that definition is a little too vague to be of use to building professionals who need to quantify actions that help meet such goals. How do we define the needs of the present when it comes to energy? How do we predict what future needs will be, if that is even possible? The term was not in use at the time (1930-1970), but one of my grandfathers practiced "sustainable farming" when he turned cow manure into liquid fertilizer to fertilize his corn field. Another grandfather practiced "energy conservation" when he went around the house and made sure anything electrical was turned off when it wasn't being used. Some sustainability is common sense mixed with frugality born from necessity, but we need better than that here. In the foundations course I teach on environmental sustainability at Southern Methodist University, I define sustainable this way with an eye towards energetics and resource efficiency:

> *Sustainable means any process which is regenerative*
> *and whose goal is zero waste.*

So if we make something more energy efficient, do we also make it more sustainable? Not necessarily. If I have a coal-fired electric power plant which is roughly 40% efficient and I add cogeneration to the plant which increases the efficiency to roughly 60% (or greater), have I made the process sustainable? By our definition, generating electricity from coal is inherently unsustainable because extracting and combusting coal, since it is a finite resource, is ultimately non-regenerative.

ENERGY EFFICIENT

Anyone who has studied just a little physics, engineering or building science knows that "energy-efficient" is a relative term. Energy transformations can involve waste in various forms—heat, gases, solids—that may not end up as useful energy output. There are theoretical maximums to the efficiency of various processes; e.g., Figure 2-1 depicts efficiencies and costs related to three different generations of solar photovoltaic panels[2]:

The Lawrence Berkeley National Laboratory[3] defines the term thus, a pretty good working definition:

Energy efficiency is "using less energy to provide the same service."

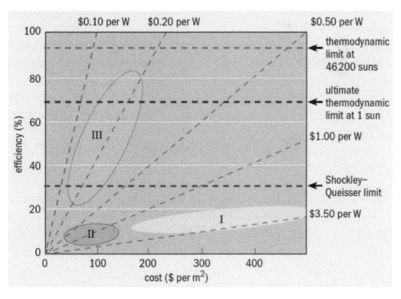

Figure 2-1

This is to be contrasted with energy conservation. Energy conservation is reducing or going without a service to save energy. An important distinction.

We can further refine the definition like this:

Energy efficient means increasing the energy output of a process by reducing the energy input and the waste in the process.

This refers to a concept in the field of energetics, the so-called EROI (or EROEI), the *energy return on energy invested*, which is a ratio of the energy generated to the energy supplied. I am much indebted in my work and teaching to Professor Vaclav Smil of the University of Manitoba and his outstanding work: *Energy in Nature and Society: General Energetics of Complex Systems.* As an approach, EROI is not without controversy as a tool for energy analysis since like all analysis tools it must ignore some things, but it is a useful concept since it allows us to look at energy efficiency from a standpoint other than financial return on investment.

"High-performance," as a general term, is intended to distinguish buildings which are designed (or retrofitted) to a higher standard of performance than is currently required or meets some percentage benchmark above the norm. *It also implies that the actual energy performance of the building must be measured and analyzed after construction, whether new or retrofitted, to verify predicted performance goals.* Nevertheless, the definition has been evolving over the last several years. What was high-performance 5 years ago may not be high-performance today, since the baseline for minimum performance goals is increasing with changes in the building codes.

For example, The US Energy Policy Act of 2005[4], Sec. 914 Building Standard, defines it as follows:

(a) DEFINITION OF HIGH-PERFORMANCE BUILDINGS. — In this section, the term "high performance building" means a building that integrates and optimizes all major high-performance attributes including energy efficiency, durability, life-cycle performance, and occupant productivity.

Two years later, the US Energy Independence and Security Act (EISA) of 2007[5], provides this definition:

Title IV—Sec. 401. DEFINITION. The term "high-performance build-ing" means a building that integrates and optimizes on a life-cycle basis all ma-jor high-performance attributes, including energy conservation, environment, safety, security, durability, cost-benefit, productivity, sustainability, function-ality, and operational considerations.

This, of course, is a definition provided by the federal govern-ment for government buildings and is intended to be all-inclusive, even going so far as to broaden the definition to include categories such as safety (which requirements are already mandated by code, or may be concerned with terrorist attacks) and productivity (somewhat difficult to track on a comparative basis). Some of the categories mentioned may be difficult to quantify or do not already have metrics to do so and I will refer to this as the "kitchen sink" definition, one that we will not use in this book.

Recently, a High-Performance Buildings Congressional Caucus Coalition was formed, and there is now a *High Performing Buildings* on-line publication from the American Society of Heating, Refrigeration & Air-Conditioning Engineers (ASHRAE)[6]. ASHRAE also now offers its own Certification in High-Performance Building with collaboration from the Illuminating Engineering Society of North America (IESNA), the Mechanical Contractors Association of America (MCAA), the Unit-ed States Green Building Council (USGBC), the Green Building Initia-tive (GBI), etc. A recent book entitled *High-Performance Building*, written ostensibly for the architectural community, states this:

Buildings designed for sustainability in the 21st century should draw on natural resources responsibly, and should provide a comfortable environment for their users. Any building that claims to be recognized as great architec-ture should also qualify as a high-performance building in terms of energy efficiency.[7]

The National Institute of Building Science is promoting their "Whole Building Design Guide" which emphasizes the primacy of high-performance building.[8] All of this means that there is consider-able support for the category and a push to make it part of the common language of the industry and part of the prescriptive *and performance* re-quirements in design, engineering and construction. It is also clear that the overall thrust is to unify different objectives with high performance to provide an integrated systems approach.

An integrated, whole-building systems approach requires that all participants in the process—design, engineering, construction, owners, lenders—"get in the same room." The evolution of high-performance building practice is closely tied to the *design-build process* and has roots in "concurrent engineering" in manufacturing. Without this approach, the efficiency of the design process and the ultimate performance of the building can be compromised by recurrent re-designs generated by unclear or incomplete objectives. The whole team must be open to continuous informational inputs during the process regarding capital investment, structure, energy efficiency, design, occupant comfort, operational and maintenance parameters, etc.

In some respects concurrent engineering is an old concept harking back to techniques developed by Henry Ford and Edward Deming, but the most current definition was given by the Institute for Defense Analysis in report R-339 which developed the actual term "concurrent engineering" in 1986:

Concurrent engineering is a systematic approach to the integrated, concurrent design of products and their related processes, including manufacture and support. This approach is intended to cause the developers, from the outset, to consider all elements of the product life cycle from concept through disposal, including quality, cost, schedule, and user requirements.[9]

Consider Figure 2-2[10] comparing the traditional, or "waterfall" approach on the left with concurrent engineering, the "iterative approach":

There are some basic rules of concurrent engineering which can be framed as follows:

- Create multifunctional design teams
- Improve communication with the customer/user
- Design processes concurrently with the product (building)
- Involve suppliers and subcontractors early
- Simulate product (building) performance
- Simulate process performance
- Integrate technical reviews
- Integrate CAE (BEA—building energy analysis) tools with the product (building) model
- Continuously improve the design process[11]

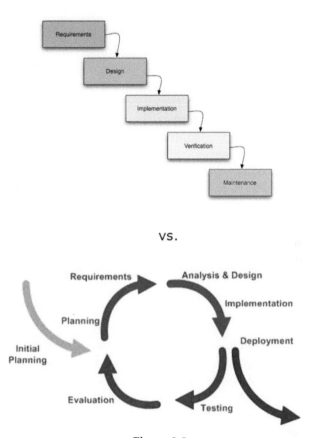

Figure 2-2

The same principles can obviously be applied to the design, engineering, construction, operation and maintenance of high-performance buildings, and in fact it must be stated that *high-performance building cannot be achieved without the iterative, concurrent engineering approach.*

We are now in a position to provide an operational definition of the subject of this book:

High-performance building is an integrated systems approach to design, engineering, construction and operations which cuts waste, optimizes resource efficiency, improves cost, reduces environmental impact and maximizes occupant comfort.

CONCLUSIONS

It is clear that high-performance building requires a shift in the way we think about energy use, buildings, and also, Return on Investment. It is not within the scope of this book to do analysis of the potential financial returns of high-performance buildings. But we can make some general statements bearing on the subject even as problematic as that is with extraordinary volatility in the cost of capital, the price of real estate and the price of energy which is now commonplace.

In the field of Energetics one talks about *energy flows* and *energy stores*. Energy flows are energy resources such as solar energy, wind energy, geothermal energy, marine energy, gravitational energy. In other words, sources of energy which provide a constant flux. Energy stores are materials which have energy stored or embedded in them, such as coal, natural gas, oil, uranium. Aside from debates about the *technically recoverable resources*[12] of these latter, the fact is that they are finite, not replaceable and can be considered the natural capital of the earth. To use a financial term, they constitute the *fixed energy capital* of the planet. When we extract and combust this fixed energy capital, we are releasing the embedded energy in the form of *interest on the capital*.[13] But once released, the capital is consumed and we cannot recover the interest on the capital or the capital itself. And as the resource becomes increasingly scarce, the financial and energetic costs required to obtain the capital and release the interest rises.

Buildings incur extensive energy and financial costs for their construction, operation, demolition, recycling, and for our purposes they have to be considered long-term assets. The stocks of existing buildings are major repositories of *embedded energy as well as asset value*. Although the recycling of building materials has become increasingly sophisticated and efficient at capturing energy and applications for reuse, recycling processes require significant energy inputs, typically from fossil fuel sources.

So high-performance building has to have a major focus on the reuse and repurposing of existing buildings as well as the design and construction of new buildings. This contributes to the reduction in consumption of finite sources of energy and the mitigation of environmental impact.

In real estate and construction, there is always a trade-off between short-term capital efficiency and long-term resource management. If we want buildings to fulfill the goals of high performance, then we will have to be more patient about the financial return on the money invested. We will have to build buildings again as if we expect them to last a hundred years.

References

1. Quoted in: Introduction to Architectural Science, Steven V Szokolay, Architectural Press 2004, p. 234
2. Wikipedia commons: solar energy. Generation I: crystalline silicon, Generation II: CIGS, amorphous thin-film, Generation III: quantum dot, tandem/multijunction, polymer cells, etc.
3. http://www.eetd.lbl.gov/ee/ee-1.htm
4. http://www.doi.net/iepa/EnergyPolicyActof2005.pdf
5. http://www.energy.senate.gov/public/_files/RL342941.pdf
6. http://www.hpbmagazine.org/
7. High-Performance Building, Vidar Lerum, John Wiley 2008, p. 6.
8. http://www.wbdg.org
9. Quoted in: http://www.lukerickert.com/THE_HISTORY_OF_CE5.pdf
10. Wikipedia commons: concurrent engineering
11. Managing Concurrent Engineering, Turino, J., Van Nostrand Reinhold, 1992
12. "Technically recoverable resources are those amounts within the basic resource which are deemed to be recoverable using current technology, regardless of whether this would be justifiable economically." For an interesting discussion about the meaning of reserves and resources, see: Energy, Dukert, J., Greenwood Press, 2009, chapter 2.
13. See: Matter and Energy, Frederick Soddy, Henry Holt & Company, 1912, p. 139.

Chapter 3

High Performance—
A Real Estate Perspective

Jyl DeHaven, MS
Tony Robinson, MS

INTRODUCTION

The real estate industry makes the majority of their investment decisions based on ROI calculations. It is an industry that has been challenged to quantify and articulate the value of sustainable property investment. In order to incorporate sustainability and integrate the benefits beyond cost savings into their valuations and underwriting requires a reassessment of how to calculate the performance of properties.

Beginning in the 1960s, we became very good at creating built environments that did not have to have any significant long-term value. The buildings of our nation reflected a social change towards increasing convenience and instant gratification. We built buildings that didn't have to survive the life of the loan. Somewhere along the line after WWII, quality was replaced with quantity, and time and value became disconnected. We wanted buildings that were completed in the shortest period of time with the highest return on our money. How they performed long-term after completion, both economically and operationally, became a lower priority. Since buildings were rarely held long-term by the original developer, decisions that occurred during the design and construction phase were about ROI at the time of the sale of the property. Long-term performance and efficiency became "someone else's" problem.

Now as we sit at the beginning of the second decade of the 21st century, we have a huge stockpile of low performing, high energy using, high waste producing buildings and a market that is sluggish on new construction. The building industry is definitely at a fork in the road. What do we do now? Is sustainability or "green" the silver bullet everyone is searching for? Can we afford to stand still and hope the "good old days" will return? The answers to the last two questions are NO and NO. There is no silver bullet, but to think that business as usual is going to return is the wrong "fork" to take. Sustainable business practices are the strongest tools that we have available to "fix" the mess that we are in, and high-performance buildings can provide a fundamentally important direction for the real estate industry.

There is now a growing understanding by many investors and tenants that sustainable properties can generate health and productivity benefits, recruiting and retention advantages, and risk reduction. However, to actually place a monetary value on these improvements demands a paradigm shift from "business as usual." This requires the ability of property investors to appropriately incorporate revenue and risk management decisions into sustainable investment decisions.

With increasing government regulations, incentives, and rapidly growing tenant and investor interest in sustainability, the real estate industry needs to incorporate value considerations beyond cost savings, or it will eventually see sub-optimal financial results for investors.[1]

Overall, there are five basic areas of real estate investing: *build, buy, operate, lease and finance.* Each of these areas requires different criteria when it comes to assessing sustainability. There are numerous sustainable certifications, features, and available performance data that can be used to drive sustainable decisions, but each building's sustainable menu will vary significantly, so evaluating sustainable property requires sophisticated analysis and that the owner be very diligent with his/her own homework.

Sustainability is a combination of features rather than a property "type," so underwriting, modeling and data collection seldom reflect entirely the true value of the property. This data gap adds to the challenge of identifying and justifying true value-add standards for investment and financing purposes. Even geographic location changes the equation. Until the "roadmap" for sustainability is developed and historical performance data are captured long-term, investors, property managers, and builders must develop some of their own "knowledge economy" for making the best decisions.

So how do you as a building owner or property manager determine the potential value of high-performance/sustainable building? If you are looking at the possible purchase of a "green" building, you can review the past performance—both from an operational and financial perspective—and extrapolate to high-performance in all of the line items. Better yet, you can also interview tenants and facilities managers to see what their actual experience has been with the building. Was there a noticeable improvement in comfort, productivity and fewer sick days? Fewer hot calls? Lower maintenance? All of this information has a value and should be included in the evaluation model. If you own/ manage a building, it is recommended that you develop a system for capturing this information for future use. This data collection can be done electronically, through various monitoring systems, as well as with a traditional questionnaire/written documentation. It may be of value to look to outside industry consultants to be part of your team to accomplish this research. Although the industry has generated and captured some historical performance data, you as the owner/manager of a building will have to take on some of the responsibility to collect the data and contribute to the growing body of knowledge about high-performance buildings. The good news is that the value you will be able to show will equate to financial benefits in both the short and the long term.

FRAMEWORKS

So let's look more closely at the business case for sustainability. The conceptual definition of sustainability is three-cornered, a Venn Diagram: environmental responsibility, social responsibility, and economic viability. Where economic and environment overlap—*viable*. social and environmental overlap equals *bearable*, and social and eco-

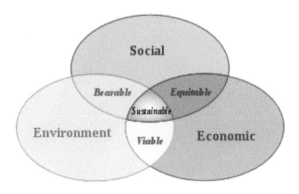

Figure 3-1

nomic is *equitable*. *Sustainable*, in the center, is the intersection of all of these sets.

Historically, in the green-building industry, we have focused on the first two categories—environmental and social responsibility, but have done a rather poor job of educating on the economics of sustainability. Most owners of real estate have seen a big separation between "do good" and "make money." The perception is that making a building "green," perhaps LEED certified, has added an expense that often could not be justified. Only long holds of real estate—primarily institutions and municipalities—could make the numbers work. The private sector merely looked at sustainability or high performance as a luxury that made no economic sense.

The good news is that time and energy costs are changing that equation. The transformation of business as usual brings into the spotlight that doing good can make money and making money can be done while doing good. In fact, the closer you can get those two phrases to link—the stronger the business case becomes. Look around at Fortune 500 businesses and how the word "sustainability" is now on almost all

What Is Green Building?

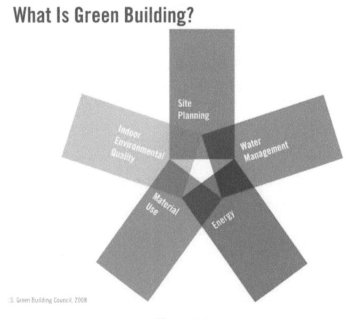

Figure 3-2

of their websites. And see below some of the substantial growth in investment opportunities in this sector.

More specifically, we have numerous frameworks for evaluation to look at, including, although not limited to, one of the most recognizable from the United States Green Building Council (USGBC).

Now it is true that the priorities within the "green building" movement have been, at least initially, to look at better ways to address environmental concerns, with a slow evolution towards social responsibility. And the sustainable "third leg" called economics has often gotten lost along the way. For a little while there, the purveyors of "green" were sure that building more energy efficient buildings made good economic sense, but what did that equation really look like? We forgot that without the business case for sustainability, the movement towards high performance would get left mainly to government institutions and municipalities. But the bottom line isn't just about being a good citizen, it's about cutting waste and saving money too.

Immediate Advantages of Green Buildings
Lower operating costs
Improved indoor air quality
Increased value

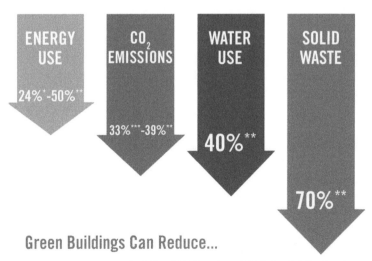

Green Buildings Can Reduce...

* Turner, C. & Frankel, M. (2008). Energy performance of LEED for New Construction buildings. Final report.
** Kats, G. (2003). The Costs and Financial Benefits of Green Building: A Report to California's Sustainable Building Task Force.
*** GSA Public Buildings Service (2008). Assessing green building performance: A post occupancy evaluation of 12 GSA buildings.

Figure 3-3

Reduced liability
Improved risk management, includes enhanced productivity
Enhanced recruitment
Reduced absenteeism
Governmental incentives

Long-term Advantages of Green Buildings
Creation of green-collar jobs
Improved air quality
Improved public health
Reduced urban heat island effect
Reduced peak demand for electricity
Reduced utility costs for businesses and consumers
Economic development opportunities
Reduced demand for potable water

So, what if you could start to look at your building(s) as assets, whether they have tenants or not? We have created a real estate market that has built up the economic value of buildings on how many people they can hold and how well they do that. We have also developed a marketplace where the value of the building has very little to do with cost and a great deal to do with how fast we make the money move. Somewhere along the way, we have forgotten to look at the building itself as something that holds long-term value and generates benefits beyond its walls as well as income-producing benefits for the owners.

This chapter could also be called "slow money down, so buildings can perform long-term."

MARKET DEVELOPMENT & ACCEPTANCE

The market for sustainability and high-performance has been growing steadily for the last five years and shows no sign of backing off. Consider the following:

USGBC North Texas Chapter[2]
Local government activity from the US Conference of Mayors:
−*75% of cities* are changing building codes and/or ordinances to encourage green or sustainable buildings.

—88% of cities are educating the public about the need for buildings which are energy efficient, healthy and environmentally sustainable.

US Social Investment Forum[3]

Estimates *$3 Trillion* of investments using environmental, social, and governance criteria (ESG) 12.2% of all investment assets under management in the USA—up 34% from 2005-2010

Investor Network on Climate Risk (CERES)[4]

—Estimates *$1.2 trillion* in green investment in the pipeline.

CoStar Study[5]

The CoStar Group in association with CB Richard Ellis and the US-GBC, have published numerous studies and done valuable research on the connection between green building, tenant preference, comfort and productivity and investor returns. Some conclusions from a 2011 study as follows:

- *Green buildings demonstrate higher average occupancy levels than the general market.*

- *LEED and Energy Star buildings show higher rental rates. Non-LEED buildings had 4.81% lower average rents than the broader market while LEED buildings were 7.38% higher.*

- *Nearly half of commercial building owners expect to see heightened green building activity in three years compared to 2010. "There is momentum in the marketplace, at least among the institutional owners we represent," Pogue (Director of Sustainability for CBRE) said.*

- *Institutional owners are mostly investing in green practices with the expectation of an economic payoff, with 93% of those surveyed anticipating a decrease in operating costs, while 79% and 73% expected to attract more tenants and receive better building ROI, respectively.*

- *Just under 80% of owners agree or strongly agree that LEED status can be achieved affordably with low-cost features. Pogue estimated the average cost to upgrade existing buildings to LEED (EBOM) of the 100 or so buildings CBRE has managed at about 25 cents per square foot.*

- *Owners are driven most by holistic business benefits rather than looming government regulation or social goals, with nearly 80% citing energy reduction and 72% citing competitive advantages from offering "green" features. Building managers are driven overwhelmingly by cost savings. There is no tree hugging among property managers.*

A survey in May of 2012 from CoStar also revealed the following about rentable building area (RBA)[6]:

RBA in LEED-Certified Office Space, 2006-2011

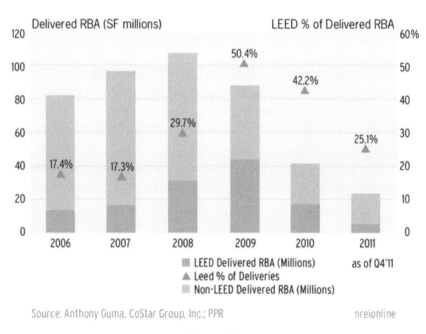

Source: Anthony Guma, CoStar Group, Inc.; PPR   nreionline

Figure 3-3

BOMA

For the past five years The Building Owner's and Managers Association (BOMA) has issued its *Market Transformation Energy Plan & 7-Point Challenge*[7] to its members which has been well accepted and includes the following:

The 7-Point Challenge
1.   *Continue to work towards a goal to decrease energy consumption by 30 percent across your portfolios by 2012 — as measured against an*

"average building" measuring a 50 on the ENERGY STAR® bench-marking tool in 2007.

2. *At least once a year, benchmark your energy performance and water usage through EPA's ENERGY STAR benchmarking tool (and share your results with BOMA);*

3. *Provide education to your managers, engineers, and others involved in building operations, to ensure that equipment is properly maintained and utilized;*

4. *Perform an energy audit and/or retro-commissioning of your building, and implement low-risk, low-cost and cost effective strategies to improve energy efficiency with high returns;*

5. *Extend equipment life by improving the operations and maintenance of building systems and ensure equipment is operating as designed;*

6. *Through leadership, positively impact your community and your planet by helping to reduce your industry's role in global warming; and*

7. *Position yourself and the industry as leaders and solution providers to owners and tenants seeking environmental and operational excellence.*

WGBC

The World Green Building Council has recently issued its report, "The Business Case for Green Building: A Review of the Costs and Benefits for Developers, Investors and Occupants":

"This report synthesizes credible evidence from around the world on green buildings into one collective resource, and the evidence presented highlights that sustainable buildings provide tangible benefits and make clear business sense," says (sic) Jane Henley, CEO of World GBC. "From risk mitigation across a building portfolio and city-wide economic benefits, to the improved health and well-being of individual building occupants, the business case for green building will continue to evolve as markets mature. Indeed we have already seen this momentum grow globally where in more and more places, green is now becoming the status quo."[8]

So how does all of this dovetail with the marketing case for high-performance? Simply put, public perception is changing to view sustainable, high-performance buildings as more competitive, with better amenities. A more integrated view of the building as a whole system is gaining acceptance. And high performance building technology refers to the whole system, or industry, of expertise, practices, processes and products required for the planning, design, development, implementation and operation of high performance buildings. Used in this context, technology has a broad connotation referring to the processes by which an organization transforms labor, capital, materials, and information into products and services of greater value. It extends beyond engineering and manufacturing to encompass a wide range of marketing, investment, and managerial processes.

Companies and consumers are starting to view the built environment as something more than a place they live, work, or play. Their knowledge of sustainability—all three arms of that definition is growing and building owners and developers are starting to find themselves in the role of having to use more than competitive rent rates as a reason to attract tenants. Using high performance to make your product (building) different from the competition becomes a strong tool in an increasingly competitive market.

Although some of these companies might be accused of "green washing," the fact is they are beginning to accept the reality that they can no longer operate inside of their own silos, but must work collaboratively and responsibly if they want to insure long-term profitability. Moreover, sustainability should be about operational reality first, and public perceptions second. Competitive markets and the more knowledgeable tenant gives the developer/building owner the opportunity to leverage high performance buildings in order to add value, rent quicker, and maintain leases longer.

CONCLUSIONS

The adage "time is money" is evolving into "high performance is money," a fundamental "value-add." The operational savings and risk mitigations of high performance buildings are slowly finding their way into acceptable accounting practices. The currently acceptable payback timeframe for capital investments incurred on high performance buildings is less than 5 years and many high-performance measures such as lighting retrofits, monitoring systems and HVAC upgrades have less than 3 year returns. Paybacks on distributed renewable energy generation and renewable energy systems are beginning to fall into the under 10-year range when aggregated with other building improvements and federal, state and local incentives.

In addition to the governmental high-performance operational savings for the GSA and other large branches, it also no surprise that operational savings is one of the primary foci of the military. The military is the most energy-intensive branch of the government and there *the high-performance-sustainable approach also becomes a matter of national security.* I draw your attention to the following quotation from an article by someone who has "been there":

> *"It is common knowledge that the Department of Defense (DOD) has considered the science of climate change and it has accepted the conclusion that it is real (reference the 2010 Quadrennial Defense Review1). Climate change and energy efficiency are now standing military planning considerations as both will impact the 21st Century strategic landscape and operational environment to such a degree that to not address them would be a gross dereliction of duty and an abrogation of responsibility2.*
>
> *As such, plans are being developed and executed to address the risks of climate change to military installations and global military operations. In addition, all the services within the DOD have acknowledged and are addressing the operational and tactical necessity of developing deployable renewable energy technologies to achieve energy autonomy in the field as a means of reducing logistical vulnerabilities3. Simply put, within the U.S. military, there is no debate about the risks, threats, and challenges of climate change and energy dependency as they relate to our national security. Accordingly, the U.S. military is taking prudent action both at home and abroad to address these issues.*
>
> *I hope highlighting the fact that the military takes climate change seriously helps with the national discourse amidst a great deal of craziness on the subject. But there is a more important issue underlying all of our national anxieties and fears. We've become too ready to cast off our responsibilities as citizens by outsourcing them to other segments or institutions of our society. Nowhere is this more prevalent than when it comes to "national security." Whether it is the rise of China, Iranian saber rattling, global terrorism, the war on drugs, illegal immigration, hurricanes, earthquakes, nation building, or now climate change and renewable energy, our country*

has become too comfortable in looking to the military in particular to address every vexing issue that emerges in our lives.

That's why I want to talk about national security and the responsibility that comes with citizenship. Specifically, I want to challenge my fellow citizens to take their security into their own very capable hands and, in the process, shape the course and destiny of our nation. The fact of the matter is that security in the 21ˢᵗ Century has little to do with what security was all about in the 20ᵗʰ Century. National security, as it is defined in the 21ˢᵗ Century, can no longer be considered just a military issue that begins at our shores and extends outward. 21ˢᵗ Century security is more about the vibrancy and resilience of the essential systems that operate within our borders and that are so intimately intertwined with the larger "global system" that constitutes human civilization, as we understand it—food, water, energy, education, industry, mobility, information, the built environment, public health, and the global ecology."[9]

If the US military takes the issues of high-performance and sustainability that seriously, it is high time the rest of us did too. So in conclusion, I would state the case from the real estate perspective as follows:

A high performance sustainable building is one that collectively generates a high return on investment for the environment, for the building owners and managers, for the occupants, and for the community.

References

1. Value Beyond Cost Savings: How to Underwrite Sustainable Properties, Green Building Finance Consortium, Muldavin, S., 2010. see: www.greenbuildingfc.com
2. http://www.northtexasgreencouncil.org/
3. http://ussif.org/
4. http://www.ceres.org/incr/
5. http://www.costar.com/News/Article/Case-for-Green-Buildings-Grows-Stronger-for-Owners-Occupants/127092
6. http://nreionline.com/brokernews/greenbuildingnews/tenants_favor_leed_space_05142012/
7. http://www.boma.org/getinvolved/7pointchallenge/Pages/default.aspx
8. http://www.ecobuildingpulse.com/news/2013/ecobuilding-pulse/new-study-highlights-financial-benefits-of-green-buildings.aspx?utm_source=newsletter&utm_content=jump&utm_medium=email&utm_campaign=EBP_030713&day=2013-03-07
9. From: National Security, Sustainability, and Citizenship, by Mark Mykleby, Colonel, USMC (Retired)

Chapter 4

Building Energy Analysis

Dr. Victoria Chen
Dr. Pin Kung
Tony Robinson, MS

Building energy analysis (BEA) is the study of energy transfers in a building, including all of those due to climate, orientation, envelope, mechanical and electrical equipment and occupant behavior. The performance characteristics of building components as well as the whole building can be simulated and the costs of energy inputs and energy modeling can be done for either new or existing buildings. There are essentially two kinds of BEA, which we will define as (1) discrete, having to do with the analysis of building components or component systems, and (2) integrated, which considers whole buildings. BEA is essential to model the design of new buildings and to evaluate their ability to meet the goals of high-performance. It is also important to do BEA for existing buildings to consider modifications or retrofits which could bring them into line with high-performance codes or standards. The US Department of Energy's Building Energy Software Tools Directory[1] lists more than 400 BEA analysis programs some of which are web-based and many of which are available at no cost. We will consider only a few.

Conducting BEA is a complicated process requiring people who are not only experienced in computer simulations but who ideally also have extensive knowledge of and experience with building systems. For this reason, BEA is best handled by a team of people who can assess the data inputs and check for errors or misinterpretations. Even under optimal conditions, with highly accurate information and experienced team members, BEA alone does not guarantee results. There are always some variables which are difficult to control or predict when it comes to the operation of buildings: climate variability, energy costs, occupant behavior, etc. This is why monitoring systems are ultimately essential to verify predicted results, quantify behaviors and provide feedback loops.

Although they do not analyze energy transfers, we will also consider some life-cycle analysis (LCA) software tools in this chapter as well since they bear on the overall concept of high-performance from the standpoint of environmental impact.

ENERGY AUDITS

Before we review some major software tools available for BEA and LCA, it is first necessary to discuss the role of energy audits. An existing building should have an energy audit before engaging BEA to verify existing conditions or before attempting any modifications to existing facilities. This establishes the building's baseline and insures the accuracy of inputs to the simulation. It is important to be aware of the extent of the information in an energy audit as this information is typically required for input into BEA simulations.

The term energy audit is commonly used to describe a broad spectrum of energy studies ranging from a quick walk-through of a facility to identify major problem areas to a comprehensive analysis of the implications of alternative energy efficiency measures sufficient to satisfy the financial criteria of sophisticated investors. Numerous audit procedures have been developed for commercial buildings, ASHRAE having developed some of the most well-known.[2]

An energy audit is required to identify the most efficient and cost-effective energy conservation opportunities (ECOs), energy conservation measures (ECMs) or energy efficiency measures (EEM). Some of the main areas of an audit consist of:[3]

- *The analysis of building and utility data, including study of the installed equipment and analysis of energy bills;*
- *The survey of the real operating conditions;*
- *The understanding of the building behavior and of the interactions with weather, occupancy and operating schedules;*
- *The selection and the evaluation of energy conservation measures;*
- *The estimation of energy saving potential;*
- *The identification of customer concerns and needs*

ASHRAE diagrams the relationship and the stages of an audit in Figure 4-1:[4]

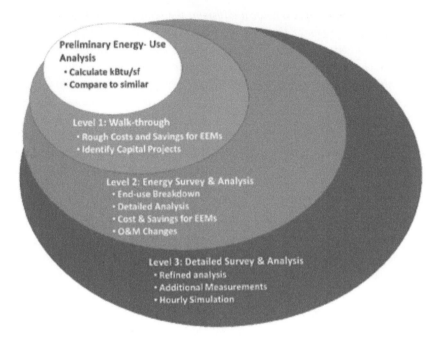

Figure 4-1

Notice that "hourly simulation"[5] is shown as the last step in level 3, indicating that BEA can be the final stage in an energy audit.

ASHRAE further outlines the audit processes as such:[6]

- *Level 0—Benchmarking: This first analysis consists of a preliminary whole-building energy use (WBEU) analysis based on the analysis of the historic utility use and costs and the comparison of the performances of the buildings to those of similar buildings. This benchmarking of the studied installation allows determining if further analysis is required.*

- *Level I—Walk-through audit: Preliminary analysis made to assess building energy efficiency to identify not only simple and low-cost improvements but also a list of ECMs, ECOs, or EEMs to orient the future detailed audit. This inspection is based on visual verifications, study of installed equipment and operating data and detailed analysis of recorded energy consumption collected during the benchmarking phase.*

- *Level II—Detailed/general energy audit: Based on the results of the pre-audit, this type of energy audit consists in energy use survey in order to provide a comprehensive analysis of the studied installation, a more detailed analysis of the facility, a breakdown of the energy use and a first quantitative evaluation of the ECOs, ECMs or EEMs selected to correct the defects or improve the existing installation. This level of analysis can involve advanced on-site measurements and sophisticated computer based simulation tools to evaluate precisely the selected energy retrofits.*

- *Level III—Investment-grade audit: Detailed analysis of capital-intensive modifications focusing on potential costly ECOs requiring rigorous engineering study.*

Energy audits can require generalized information and/or detailed formats leading to extensive reporting for the purposes of cost estimating and capital investment review. Forms 4-1 and 4-2, following, are examples of typical data input forms.

DISCRETE BEA

Discrete BEA includes analysis tools which typically analyze components of buildings or component systems. HVAC, wall systems or windows would be examples. We shall consider here two of the more prominent applications.

TRNSYS

TRNSYS (transient system simulation program) has been used for over thirty years for HVAC analysis and sizing, multizone airflow analyses, electric power simulation, solar design, building thermal performance, analysis of control schemes. It as an older application originally written in Fortran which has been continually updated and improved.[10]

TRNSYS is a transient systems simulation program with a modular structure. It recognizes a system description language in which the user specifies the components that constitute the system and the manner in which they are connected. The TRNSYS library includes many of the components commonly found in thermal and electrical energy systems, as well as component routines to handle input of weather data or other time-dependent forcing functions and output of simulation results. The modular nature of TRNSYS

Form 4-1. Building Information (Sample Form)[8]

On the following page, prepare a site sketch of your building or building complex which shows the following information:
1. Relative location and outline of the building(s).
2. Building Age
3. Building Number (Assign numbers if buildings are not already numbered.)
4. Building Size
5. Fuel Type
6. Location of heating and cooling units
7. Heating plants
8. Central cooling system, etc.
9. North orientation arrow

2. BUILDING CHARACTERISTICS

   a. **Gross Floor Area:** _____ Gross Sq.Ft. x Ceiling Height _____ Ft. = volume ____Cu.Ft.
   b. **Conditioned Floor Area:** _____ (if different that gross floor area)
   c. **Total door Area:** _____ Sq.Ft. Glass doors _____sq.ft. Wood doors _____sq.ft. Metal doors _____ sq.ft. Garage doors _____ sq.ft.
   d. **Total Exterior Glass Area:** _____sq.ft. Single Panes _____sq.ft. Double panes ____sq.ft.

   North South East West
   Total Area _____sqft _____sqft _____sqft _____sqft
   Single Pane _____sqft _____sqft _____sqft _____sqft
   Double Pane _____sqft _____sqft _____sqft _____sqft

   e. Total Exterior Wall Area: _____ **sqft Material: [ ]Masonry [ ]Wood [ ]Concrete [ ]Stucco [ ]Other**
   f. **Total Roof Area:** _____sqft Condition: [ ]Good [ ]Fair [ ]Poor
   g. **Insulation Type:** _____Roof _____Wall _____Floor
   h. **Insulation Thickness:** _____Roof _____Wall _____Floor
   i. **Metering:** Is this building individually metered for electricity? [ ]Yes [ ]No

Is this building individually metered for natural gas? [ ]Yes [ ]No
Is this building on a control boiler system with other buildings? [ ]Yes [ ]No
j. **Describe general building condition:**

gives the program tremendous flexibility, and facilitates the addition to the program of mathematical models not included in the standard TRNSYS library. TRNSYS is well suited to detailed analyses of any system whose behavior is dependent on the passage of time. TRNSYS has become reference software for researchers and engineers around the world. Main applications include: solar systems (solar thermal and photovoltaic systems), low energy buildings and HVAC systems, renewable energy systems, cogeneration and fuel cells.[11]

Form 4-2. Energy Type[9]

Energy Type	Total Annual Use	Units	Conversion Multiplier	Thousands BTU (kBtu)	Total Annual Cost ($)
Electricity		MMBtu	1000	0	
Natural Gas		therms	100	0	
Purchased Steam			0	0	
Purchased Hot Water			0	0	
Purchased Chilled Water			0	0	
Oil #_____			0	0	
Propane			0	0	
Coal			0	0	
Thermal - On-Site Generated			0	0	
Other			0	0	
Electricity - On-Site Generated			0	0	
Thermal or Electricity - Exported			0	0	
Total				0	0

Gross Conditioned Square Feet	
EUI (energy use intensity)	#DIV/0!
Target Finder Score	
ECI (energy cost index or $/sf)	#DIV/0!

The TRNSYS input file, including building input description, characteristics of system components, manner in which components are interconnected, and separate weather data (supplied with program) are all ASCII files. All input files can be generated with a graphical user interface known as Simulation Studio. The envelope for the TRNSYS building can be three-dimensionally created in the TRNSYS3D plugin for Google SketchUp™ and edited in the TRNBuild interface. The data included in those files can be life cycle costs; monthly summaries; annual results; histograms; plotting of desired variables.

Due to its modular approach, TRNSYS is extremely flexible for modeling a variety of energy systems in differing levels of complexity. Supplied source code and documentation provide an easy method for users to modify or add components not in the standard library; extensive documentation on component routines, including explanation, background, typical uses and governing equations; supplied time step, starting and stopping times allowing choice of modeling periods. The TRNSYS program includes a graphical interface to drag-and-drop components for creating input files (Simulation Studio), a utility for easily creating a building input file (TRNBuild), and a program for building TRNSYS-based applications for distribution to non-users (TRNEdit).

A web-based library of additional components and frequent downloadable updates are also available to users. Extensive libraries of non standard components for TRNSYS are available commercially from TRNSYS distributors. No assumptions about the building or system are made (although default information is provided) so the user must have detailed information about the building and system and enter this information into the TRNSYS interface.

Figures 4-2 through 4-4 show typical screen inputs for TRNSYS.

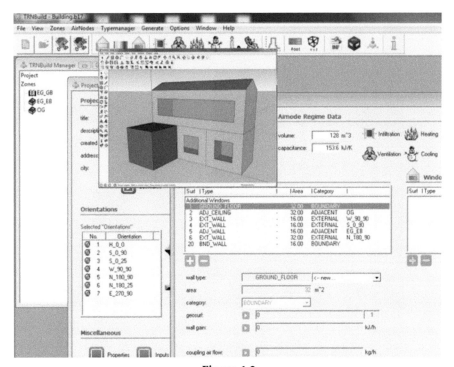

Figure 4-2

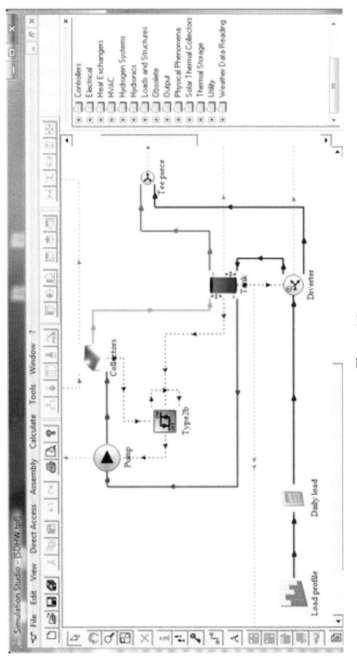

Figure 4-3

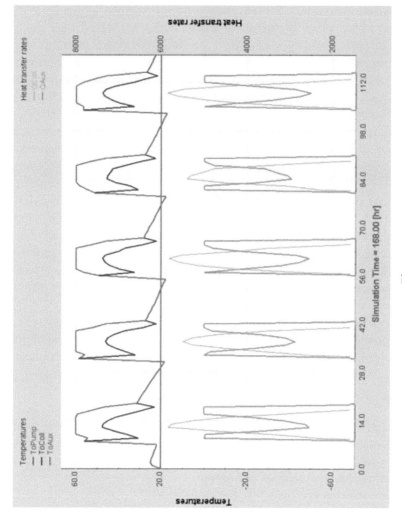

Figure 4-4

Outputs from TRNSYS can be used in other programs or as part of whole building energy analysis and have been used extensively to model renewable energy systems in on-site distributed power systems (DEG).

WINDOW-6 and THERM-6

Window-6 and THERM-6 are software programs which were developed at Lawrence Berkeley National Laboratory (LBNL) for use by manufacturers, engineers, educators, students, architects, and others to determine the thermal and solar optical properties of glazing and window systems. WINDOW 6.3 is a publicly available computer program for calculating total window thermal performance indices (i.e. U-values, solar heat gain coefficients, shading coefficients, and visible transmittances). It provides a versatile heat transfer analysis method consistent with the updated rating procedure developed by the National Fenestration Rating Council (NFRC)[12] that is consistent with the ISO 15099 standard. WINDOW6 and THERM6 implement the ISO 15099 algorithms: "Thermal performance of windows, doors and shading devices—Detailed calculations."[13] The program can be used to design and develop new products, to assist educators in teaching heat transfer through windows, and to help public officials in developing building energy codes.[14]

WINDOW-6 has the capability to model complex glazing systems, such as windows with shading systems, in particular venetian blinds. Besides a specific model for venetian blinds and diffusing layers, WINDOW-6 also includes the generic ability to model any complex layer if the transmittance and reflectance are known as a function of incoming and outgoing angles. WINDOW-6 can model single and multiple glazed fenestration products with or without solar reflective, low-emissivity coatings and suspended plastic films:

1. glazing systems with pane spacing of any width containing gases or mixtures of gases;
2. metallic or non-metallic spacers;
3. frames of any material and design;
4. fenestration products tilted at any angle;
5. shading devices; projecting products.

Two-dimensional Building Heat-transfer Modeling

THERM-6 is a state-of-the-art, Microsoft Windows™-based computer program developed at Lawrence Berkeley National Laboratory

(LBNL) for use by building component manufacturers, engineers, educators, students, architects, and others interested in heat transfer. Using THERM, you can model two-dimensional heat-transfer effects in building components such as windows, walls, foundations, roofs, and doors; appliances; and other products where thermal bridges are of concern. THERM's heat-transfer analysis allows you to evaluate a product's energy efficiency and local temperature patterns, which may relate directly to problems with condensation, moisture damage, and structural integrity. THERM's two-dimensional conduction heat-transfer analysis is based on the finite-element method, which can model the complicated geometries of building products. THERM is a module of the WINDOW-6 program. THERM's results can be used with WINDOW's center-of-glass optical and thermal models to determine total window product U-factors and solar heat gain coefficients.

Heat-transfer Analysis[15]

THERM uses two-dimensional (2D) conduction and radiation heat-transfer analysis based on the finite-element method, which can model the complicated geometries of fenestration products and other building elements. This method requires that the cross section be divided into a mesh made up of nonoverlapping elements. This process is performed automatically by THERM using the Finite Quadtree method. Once you have defined the cross section's geometry, material properties, and boundary conditions, THERM meshes the cross section, performs the heat-transfer analysis, runs an error estimation, refines the mesh if necessary, and returns the converged solution.

The results from THERM's finite-element analysis of a fenestration product or building component can be viewed as: U-factors, isotherms, color-flooded isotherms, heat-flux vector plots, color-flooded lines of constant flux, temperatures (local and average, maximum and minimum). See Figures 4-5 and 4-6.

Using WINDOW-6 and THERM-6, any glazing or fenestration system, simple or complex, can be modeled, analyzed, and the results can be used by manufacturers or inputted into whole building energy analyses.

INTEGRATED (WHOLE BUILDING) BEA

Integrated building energy analysis considers whole building systems by analyzing the energy transfers of all constituent parts using local climate data along with the costs of the energy inputs required to

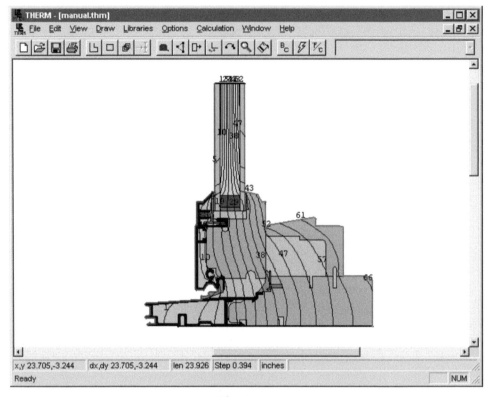

Figure 4-5

meet the specified goals. The father of all integrated BEA in the US is the computer simulation architecture developed by the US Department of Energy in the 70s based on the Fortran language and named DOE-2. There are several others in wider use at present, but it is instructive to examine the DOE architecture because the principles of subsequent software platforms are essentially the same, albeit easier to use.

DOE-2

As stated on the website, "DOE-2 is a computer simulation program for evaluating the energy performance and associated operating costs of buildings." The first version of DOE-2 was released by the Lawrence Berkeley Laboratory (LBL) in 1978 (Leighton et al., 1978). DOE-2[16] is an up-to-date, unbiased computer program that predicts the hourly energy use and energy cost of a building given hourly weather information and a description of the building and its HVAC equipment and util-

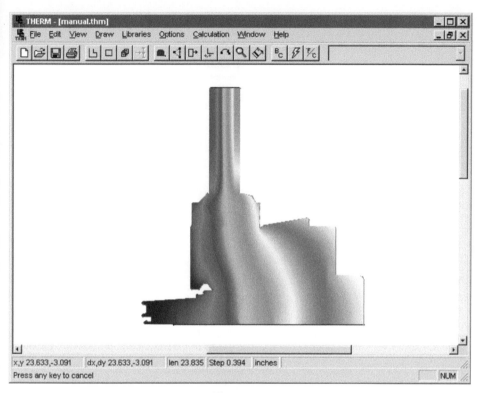

Figure 4-6

ity rate structure. Using DOE-2, designers can determine the choice of building parameters that improve energy efficiency while maintaining thermal comfort and cost-effectiveness. The purpose of DOE-2 is to aid in the analysis of energy usage in buildings; it is not intended to be the sole source of information relied upon for the design of buildings: The judgment and experience of the architect/engineer still remain the most important elements of building design.

Structure of DOE-2

Figure 4-7 shows a basic flowchart of DOE-2. DOE-2 has one subprogram for translation of the input (BDL processor), and four simulation subprograms (LOADS, SYSTEMS, PLANT and ECON). LOADS, SYSTEMS and PLANT are executed in sequence, with the output of LOADS becoming the input of SYSTEMS, etc. The output then becomes the input to ECON. Each of the simulation subprograms also produces printed reports of the results of its calculations.

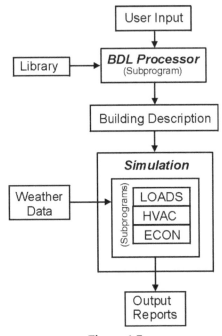

Figure 4-7

BDL Processor

The building description language (BDL) processor reads the flexibly formatted input data that you supply and translates it into computer recognizable form. It also calculates response factors for the transient heat flow in walls and weighting factors for the thermal response of building spaces.

LOADS

The LOADS simulation subprogram calculates the sensible and latent components of the hourly heating or cooling load for each user-designated space in the building, assuming that each space is kept at a constant user-specified temperature. LOADS is responsive to weather and solar conditions, to schedules of people, lighting and equipment, to infiltration, to heat transfer through walls, roofs, and windows and to the effect of building shades on solar radiation.

HVAC

The SYSTEMS subprogram handles secondary systems; PLANT handles primary systems. SYSTEMS calculates the performance of airside equipment (fans, coils, and ducts); it corrects the constant-temperature loads calculated by the LOADS subprogram by taking into account outside air requirements, hours of equipment operation, equipment control strategies, and thermostat set points. The output of SYSTEMS is air flow and coil loads. PLANT calculates the behavior of boilers, chillers, cooling towers, storage tanks, etc., in satisfying the secondary systems heating and cooling coil loads. It takes into account the part-load characteristics of the primary equipment in order to calculate the fuel and electrical demands of the building.

ECON

The ECONOMICS subprogram calculates the cost of energy. It can also be used to compare the cost-benefits of different building designs or to calculate savings for retrofits to an existing building.

Weather Data

The weather data for a location consist of hourly values of outside dry-bulb temperature, wet-bulb temperature, atmospheric pressure, wind speed and direction, cloud cover, and (in some cases) solar radiation. Weather data suitable for use in DOE-2 are produced by running the DOE-2 weather processor on raw weather files provided by the U.S. National Weather Service and other organizations.

Library

DOE-2 comes with a library of building input elements, including wall materials, layered wall constructions, and windows.

DOE-2 can be difficult to use for all but the most experienced practitioners and in the last 15 years other software platforms have been developed which provide more user-friendly interfaces.

Energy Plus[17]

EnergyPlus was introduced in 1996 and has its roots in both the BLAST and DOE–2 programs. BLAST (building loads analysis and system thermodynamics) and DOE–2 were both developed and released in the late 1970s and early 1980s as energy and load simulation tools. Their intended audience was a design engineer or architect that wished to size appropriate HVAC equipment, develop retrofit studies for life cycling cost analyses, optimize energy performance, etc. Both programs had their merits and shortcomings, their supporters and detractors, and solid user bases both nationally and internationally.

Like its parent programs, EnergyPlus is an energy analysis and thermal load simulation program. Based on a user's description of a building from the perspective of the building's physical make-up, associated mechanical systems, etc., EnergyPlus will calculate the heating and cooling loads necessary to maintain thermal control setpoints, conditions throughout a secondary HVAC system and coil loads, and the energy consumption of primary plant equipment as well as many other simulation details that are necessary to verify that the simulation is performing as the actual building would. Many of the simulation characteristics have been inherited from the legacy programs of BLAST and DOE–2.

Its architecture can be depicted as shown in Figure 4-8.

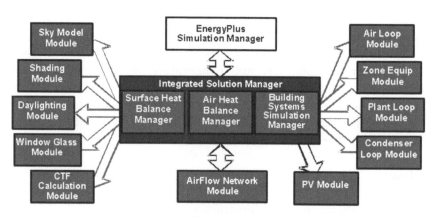

Figure 4-8

EnergyPlus is not a user interface. It is intended to be the simulation engine around which a third-party interface can be wrapped. Inputs and outputs are simple ASCII text that is decipherable but best left to a GUI (graphical user interface). This approach allows interface designers to do what they do best—produce quality tools specifically targeted toward individual markets and concerns. The availability of EnergyPlus frees up resources previously devoted to algorithm production and allows them to be redirected to interface feature development. Energy

Plus is then, for example, the simulation engine used with SketchUp Google Open Studio[18] and AutoCad's Green Building Studio.[19]

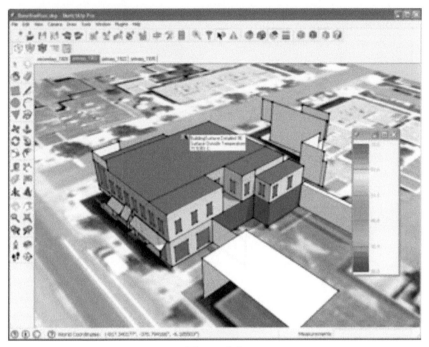

Figure 4-9

The evolution shown in Figure 4-9 quite clearly indicates the confluence of energy analysis, building design and information management into a coherent software platform.

EnergyPlus Key Capabilities[20]

The following is a representative list of EnergyPlus capabilities:

- *Integrated, simultaneous solution* where the building response and the primary and secondary systems are tightly coupled (iteration performed when necessary).

- *Sub-hourly, user-definable time steps* for the interaction between the thermal zones and the environment; variable time steps for interactions between the thermal zones and the HVAC systems (automatically varied to ensure solution stability).

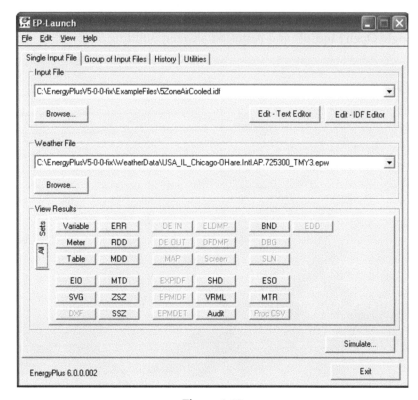

Figure 4-10

- *ASCII text based weather, input, and output files that include hourly or sub-hourly environmental conditions, and standard and user definable reports, respectively.*

- **Heat balance based solution** *technique for building thermal loads that allows for simultaneous calculation of radiant and convective effects at both the interior and exterior surface during each time step.*

- **Transient heat conduction** *through building elements such as walls, roofs, floors, etc. using conduction transfer functions.*

- **Improved ground heat transfer modeling** *through links to three-dimensional finite difference ground models and simplified analytical techniques.*

- **Combined heat and mass** *transfer model that accounts for moisture adsorption/desorption either as a layer-by-layer integration into the* *conduction transfer functions or as an effective moisture penetration depth model (EMPD).*

- *Thermal comfort models based on activity, inside dry bulb, humidity, etc.*

- *Anisotropic sky model for improved calculation of diffuse solar on tilted surfaces.*

- *Advanced fenestration calculations including controllable window blinds, electrochromic glazings, layer-by-layer heat balances that allow proper assignment of solar energy absorbed by window panes, and a performance library for numerous commercially available windows.*

- *Daylighting controls including interior illuminance calculations, glare simulation and control, luminaire controls, and the effect of reduced artificial lighting on heating and cooling.*

- *Atmospheric pollution calculations that predict CO_2, SO_x, NO_x, CO, particulate matter, and hydrocarbon production for both on site and remote energy conversion.*

Energy Plus has a web-based program called Example File Generator as well. The web-based forms allow you to enter general information about the building you want to model. The application then automatically creates a complete EnergyPlus input file, runs an annual simulation, and then emails you the EnergyPlus input, output, DXF and other files along with an annual summary of the energy results. Because EnergyPlus input files contain a lot of detail, the building description has been reduced to a set of simple, high-level parameters. OpenStudio ResultsViewer displays EnergyPlus output in a graphical format. This program provides visual representations of EnergyPlus results, making it easier to analyze the information.

The QUick Energy Simulation Tool (eQUEST)

eQUEST is a free building energy simulation tool that calculates hour-by-hour energy consumption. It combines the building energy analysis program DOE-2, graphics and three wizards, including Schematic Design (SD) Wizard, Design Development (DD) Wizards and Energy Efficiency Measures (EEM) Wizard. It is designed to provide a whole building analysis to owners, designers or operators, and building designers can use eQUEST which has the concept of integrated energy design to construct energy-efficient buildings. The building construction and operating parameters includes building envelope, internal gains, occupancy schedules, and building systems to calculate energy consumption. There are 41 wizard screens in SD Wizard. eQuest does not have quite the same scope as Energy Plus but can be easier to use since the platform is integrated with all of the necessary windows-based graphical interfaces. It is a more accessible platform than DOE-2 or Energy Plus.

An example of how it works is in the introductory tutorial version 3.63. Figure 4-11 shows the general information. In general, the 41 screens include building type, building geometry, constructions types, window sizes, door sizes, glass types, activity areas, building operations schedules, HVAC system types, power and efficiencies, water heating type, etc. Some screens are based on the previous screen options.

After SD Wizard is done, the EEM Wizard information screen is launched for simulation in Figure 4-12. The available graphical reports are shown in Figure 4-14. For example, annual building summary is

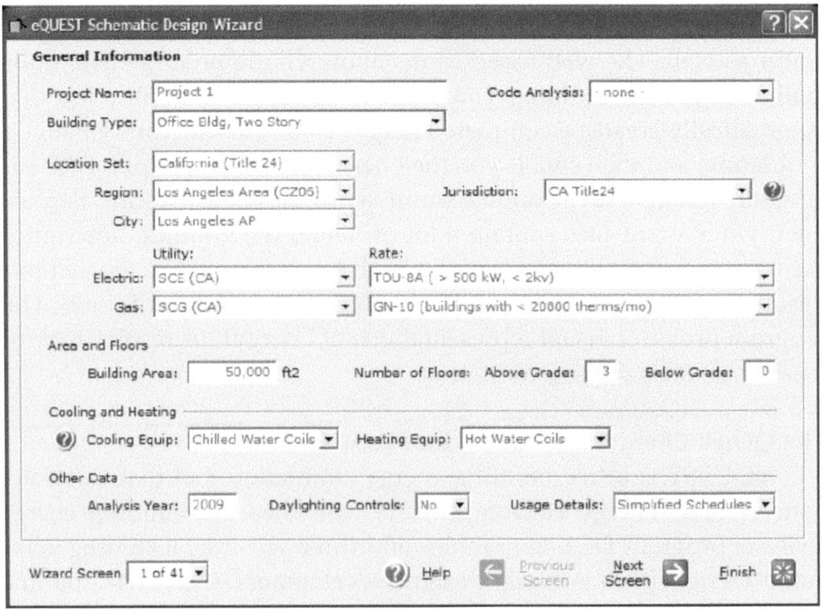

Figure 4-11. General Information

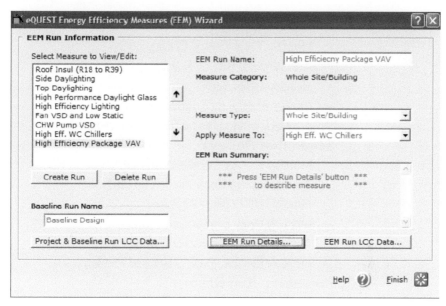

Figure 4-12. EEM Run Information

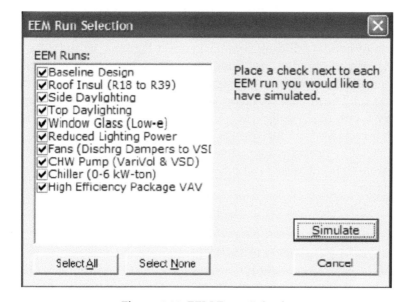

Figure 4-13. EEM Runs Selection

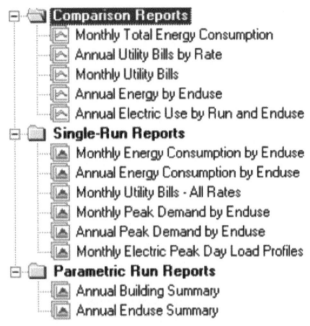

Figure 4-14. Graphical Reports

chosen from Figure 4-14 under a folder "Parametric Run Report". The reports of annual building summary are in Figure 4-15 and 4-16, and the reports are divided into three parts. The upper part has total annual results for various energy types, peak, utility cost, and life cycle cost (LCC). The middle part has incremental annual savings for various energy types, peak, utility cost, and LCC. The bottom part has cumulative annual savings for various energy types, peak, utility cost, and LCC. eQUEST provides many detailed screens for designers and energy engineers to input varied options. It is a very powerful tool for energy simulation.

LIFE-CYCLE ASSESSMENT (LCA)

Life cycle assessment (LCA) considers the stages of raw material, manufacturing, use, and end-of-life scenarios of building products and components, and it is an environmental and economic management tool which can be used to assess the life cycle of buildings.

Annual Building Summary
page 1 of 2
Annual Energy and Demand
(Parametric Report)

Annual Energy and Demand

	Ann. Source Energy		Annual Site Energy		Lighting	HVAC Energy			Peak	
	Total MBtu	EUI kBtu/sf/yr	Elect kWh	Nat Gas Therms	Electric kWh	Electric kWh	Nat Gas Therms	Total MBtu	Elect kW	Cooling Tons
Annual Energy USE or DEMAND										
0 Base Design	10,099	100.97	971,867	1,480	276,763	325,695	12	1,113	477	227
1 0+Roof Insul (R18 to R39)	10,116	101.14	973,577	1,474	276,763	327,405	6	1,118	476	226
2 1+Side Daylighting	8,888	88.87	853,636	1,479	180,533	303,684	11	1,038	428	218
3 2+Top Daylighting	8,428	84.26	806,650	1,479	144,967	294,274	11	1,005	406	215
4 3+Window Glass (Low-e)	8,291	82.90	795,344	1,475	145,930	280,018	7	950	392	195
5 4+Reduced Lighting Power	8,079	80.77	774,600	1,476	128,778	270,411	7	944	384	193
6 5+Fans (Dischrg Dampers to VSDs)	7,504	75.03	718,506	1,476	128,778	220,317	8	753	367	187
7 6+CHW Pump (VariVol & VSD)	7,137	71.36	682,659	1,476	128,778	184,470	8	630	355	187
8 7+Chiller (0-6 kW-ton)	6,832	68.31	652,860	1,476	128,778	154,670	8	529	328	187

Incremental SAVINGS	(values are relative to previous measure (% savings are relative to base case use), negative entries indicate increased use)									
1 0+Roof Insul (R18 to R39)	-17	-0.17 (-0%)	-1,709 (-0%)	6 (0%)	0 (0%)	-1,710 (-1%)	6 (49%)	-5 (-0%)	1 (0%)	1 (0%)
2 1+Side Daylighting	1,228	12.27 (12%)	119,040 (12%)	-5 (-0%)	96,230 (35%)	23,711 (7%)	-5 (-41%)	80 (7%)	49 (10%)	8 (4%)
3 2+Top Daylighting	461	4.61 (5%)	44,986 (5%)	-0 (-0%)	35,567 (37%)	9,420 (3%)	-0 (-2%)	32 (3%)	22 (5%)	3 (1%)
4 3+Window Glass (Low-e)	137	1.37 (1%)	11,306 (1%)	4 (0%)	-050 (-0%)	14,256 (4%)	4 (33%)	48 (4%)	13 (3%)	19 (9%)
5 4+Reduced Lighting Power	212	2.12 (2%)	20,744 (2%)	-0 (-0%)	17,139 (8%)	3,607 (1%)	-0 (-2%)	12 (1%)	8 (2%)	2 (1%)
6 5+Fans (Dischrg Dampers to VSDs)	574	5.74 (6%)	56,094 (6%)	-0 (-0%)	0 (0%)	56,095 (17%)	-0 (-3%)	191 (17%)	17 (4%)	7 (3%)
7 6+CHW Pump (VariVol & VSD)	367	3.67 (4%)	35,847 (4%)	0 (0%)	0 (0%)	35,847 (11%)	0 (0%)	122 (11%)	12 (2%)	0 (0%)
8 7+Chiller (0-6 kW-ton)	305	3.05 (3%)	29,800 (3%)	0 (0%)	0 (0%)	29,800 (9%)	0 (0%)	102 (9%)	29 (6%)	0 (0%)

Cumulative SAVINGS	(values (and % savings) are relative to the Base Case, negative entries indicate increased use)									
1 0+Roof Insul (R18 to R39)	-17	-0.17 (-0%)	-1,709 (-0%)	6 (0%)	0 (0%)	-1,710 (-1%)	6 (49%)	-5 (-0%)	1 (0%)	1 (0%)
2 1+Side Daylighting	1,211	12.10 (12%)	118,231 (12%)	1 (0%)	96,230 (35%)	22,001 (7%)	1 (6%)	75 (7%)	50 (10%)	9 (4%)
3 2+Top Daylighting	1,671	16.71 (17%)	163,217 (17%)	1 (0%)	131,796 (48%)	9,420 (3%)	1 (6%)	107 (10%)	72 (15%)	12 (5%)
4 3+Window Glass (Low-e)	1,808	18.08 (18%)	176,523 (18%)	5 (0%)	130,847 (47%)	45,677 (14%)	5 (39%)	156 (14%)	85 (18%)	32 (14%)
5 4+Reduced Lighting Power	2,020	20.20 (20%)	197,267 (20%)	4 (0%)	147,986 (53%)	49,284 (15%)	4 (37%)	169 (15%)	93 (19%)	34 (15%)
6 5+Fans (Dischrg Dampers to VSDs)	2,595	25.94 (26%)	253,361 (26%)	4 (0%)	147,986 (53%)	105,379 (32%)	4 (34%)	360 (32%)	110 (23%)	40 (18%)
7 6+CHW Pump (VariVol & VSD)	2,962	29.61 (29%)	289,208 (30%)	4 (0%)	147,986 (53%)	141,226 (43%)	4 (34%)	482 (43%)	122 (26%)	40 (18%)
8 7+Chiller (0-6 kW-ton)	3,267	32.66 (32%)	319,008 (33%)	4 (0%)	147,986 (53%)	171,025 (53%)	4 (34%)	584 (52%)	151 (32%)	40 (18%)

Figure 4-15. Annual Building Summary page 1 of 2

Life Cycle in Sustainable Architecture (LISA)

LISA is a simple free LCA tool for construction, to identify environmental issues and evaluate the environmental impacts resulting from building design. LCA methodologies are often too complicated and not widely accessible to designers and specifiers. Also, detailed LCA studies often divert attention from the key environmental issues, and tend to focus attention on inter-material competition, rather than on optimum construction systems.[21]

LISA includes some case studies, such as a university faculty building, multi-story offices, high rise, wide span warehouse, and road and rail bridges. Life cycle of the building in LISA contains construction, uti-

Annual Building Summary
page 2 of 2
Annual Costs
(Parametric Report)

Annual Costs

		Annual Utility Cost				Incentives		LCC
	Electric kWh($)	Electric kW($)	Electric Total($)	Nat Gas Total($)	Total ($)	Owner ($)	Design Team ($)	Total (PV$)
Annual COST								
0 Base Design	$127,081	$68,202	$199,067	$1,235	$200,922	--	--	$3,326,820
1 0→Roof Insul (R18 to R39)	$128,056	$68,217	$199,807	$1,231	$201,118	--	--	$3,330,152
2 1→Side Daylighting	$131,737	$60,327	$175,648	$1,254	$176,902	--	--	$2,926,619
3 2→Top Daylighting	$105,500	$57,411	$166,494	$1,254	$167,748	--	--	$2,774,097
4 3→Window Glass (Low-e)	$103,521	$55,739	$162,844	$1,252	$164,096	--	--	$2,713,281
5 4→Reduced Lighting Power	$100,806	$54,426	$158,817	$1,252	$160,069	--	--	$2,646,184
6 5→Fans (Dischrg Dampers to VSDs)	$93,245	$51,025	$147,854	$1,252	$149,106	--	--	$2,463,520
7 6→CHW Pump (VarVol & VSD)	$88,524	$49,242	$141,350	$1,252	$142,602	--	--	$2,355,351
8 7→Chiller (0-6 kW-ton)	$84,202	$45,200	$133,084	$1,252	$134,336	--	--	$2,217,424

Incremental SAVINGS   (values are relative to previous measure (% savings are relative to base case cost), negative entries indicate increased cost)

1 0→Roof Insul (R18 to R39)	$-185	$-15	$-200	$4	$-196	--	--	$-1,333
2 1→Side Daylighting	$16,329	$7,890	$24,219	$-3	$24,216	--	--	$403,533
3 2→Top Daylighting	$6,237	$2,916	$9,154	$0	$9,154	--	--	$152,522
4 3→Window Glass (Low-e)	$1,979	$1,672	$3,650	$2	$3,652	--	--	$60,816
5 4→Reduced Lighting Power	$2,715	$1,313	$4,027	$0	$4,027	--	--	$67,067
6 5→Fans (Dischrg Dampers to VSDs)	$7,561	$3,401	$10,963	$0	$10,963	--	--	$182,664
7 6→CHW Pump (VarVol & VSD)	$4,721	$1,783	$6,504	$0	$6,504	--	--	$108,369
8 7→Chiller (0-6 kW-ton)	$4,322	$3,943	$8,266	$0	$8,266	--	--	$137,727

Cumulative SAVINGS   (values (and % savings) are relative to the Base Case, negative entries indicate increased cost)

1 0→Roof Insul (R18 to R39)	$-185	$-15	$-200	$4	$-196	--	--	$-1,333
2 1→Side Daylighting	$16,144	$7,875	$24,019	$1	$24,020	--	--	$400,201
3 2→Top Daylighting	$22,381	$10,791	$33,173	$1	$33,174	--	--	$552,723
4 3→Window Glass (Low-e)	$24,360	$12,463	$36,823	$3	$36,826	--	--	$613,539
5 4→Reduced Lighting Power	$27,075	$13,776	$40,850	$3	$40,853	--	--	$680,636
6 5→Fans (Dischrg Dampers to VSDs)	$34,636	$17,177	$51,813	$3	$51,816	--	--	$863,300
7 6→CHW Pump (VarVol & VSD)	$39,357	$18,960	$58,317	$3	$58,320	--	--	$971,669
8 7→Chiller (0-6 kW-ton)	$43,679	$22,903	$66,583	$3	$66,586	--	--	$1,199,395

Figure 4-16. Annual Building Summary page 2 of 2

lization, refurbishment, and recycling. The user can specify the amount of materials that are recycled, and transport type distances. Figure 4-17 shows the main screen from LISA. In Figure 4-18, some case studies are shown in the open case study screen and a few descriptions from each case study appear below that. For example, Fences-Australia is chosen in Figure 4-18. The detailed information is from the right five items, such as specification, construction, repair/maintenance, decommissioning, and material transport. Clicking the "Reports" main menu displays three options in Figure 4-19. The report of an input chart is shown in Figure 4-20. Repair/Maintenance has the highest value 125, and construction has the second high value 63. Figure 4-21 shows the report:

Bill of Materials. The two stages construction and repair/maintenance display the amounts. Figure 4-22 illustrates the report of base material data. The values and recycling credits are displayed for all materials.

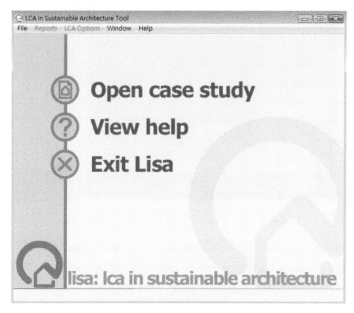

Figure 4-17. The main screen for LISA

Figure 4-18. Case Studies

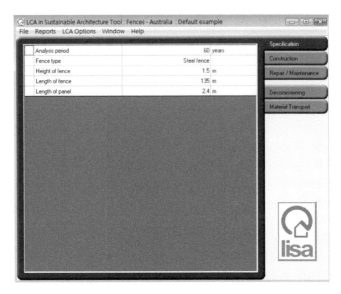

Figure 4-19. Example of Fences—Australia

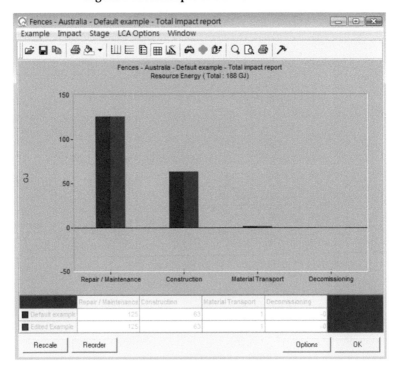

Figure 4-20. An impact chart

Stage	Item	Component	Material	Amount	Unit
Construction	Steel fence type	Colorbond	Concrete	1.4	m3
			Steel - powder coat	1.4	t
		Decomissioning	Concrete	-1.4	m3
		Material Transport	Road	140	tkm
Repair / Maintenan	Colorbond fence life		Concrete	2.7	m3
			Steel - powder coat	2.7	t
		Decomissioning	Concrete	-2.7	m3
		Material Transport	Road	280	tkm

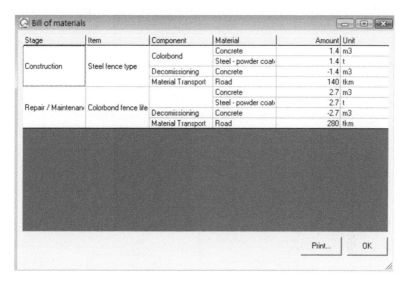

Figure 4-21. Bill of Materials

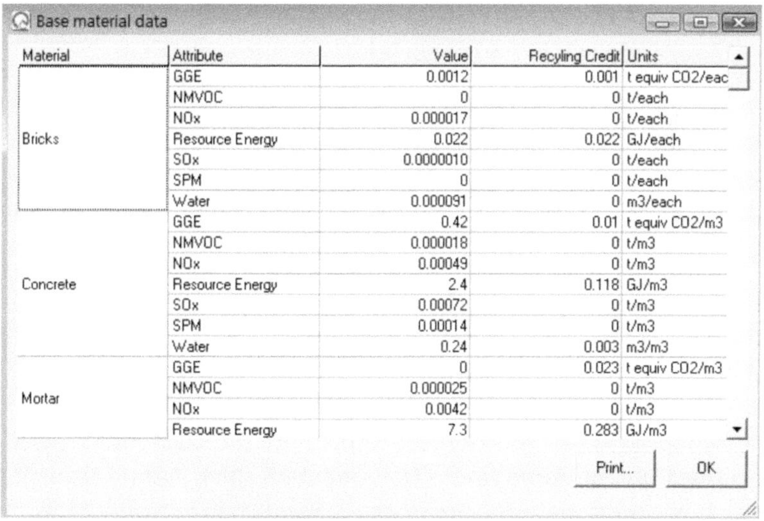

Material	Attribute	Value	Recyling Credit	Units
Bricks	GGE	0.0012	0.001	t equiv CO2/eac
	NMVOC	0	0	t/each
	NOx	0.000017	0	t/each
	Resource Energy	0.022	0.022	GJ/each
	SOx	0.0000010	0	t/each
	SPM	0	0	t/each
	Water	0.000091	0	m3/each
Concrete	GGE	0.42	0.01	t equiv CO2/m3
	NMVOC	0.000018	0	t/m3
	NOx	0.00049	0	t/m3
	Resource Energy	2.4	0.118	GJ/m3
	SOx	0.00072	0	t/m3
	SPM	0.00014	0	t/m3
	Water	0.24	0.003	m3/m3
Mortar	GGE	0	0.023	t equiv CO2/m3
	NMVOC	0.000025	0	t/m3
	NOx	0.0042	0	t/m3
	Resource Energy	7.3	0.283	GJ/m3

Figure 4-22. Base Material Data

Building for Environmental and Economic Sustainability (BEES)

BEES is a free software which is based on a LCA approach to obtain economic and environmental performance results. BEES was developed by the NIST and measures the environmental performance of building products by using the life-cycle assessment approach specified in the ISO14040 series of standards. BEES product data contain raw materials, manufacturing, transportation, installation, use, and end of life, and the LCC method covers the costs of initial investment, replacement, operation, maintenance and repair, and disposal. It is designed as shown in Figure 4-23.[22]

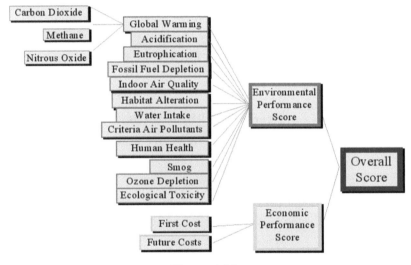

Figure 4-23

BEES also describes the entire life cycle of the building product from the tutorial in detail. An example supposes (1) Environmental performance is 50%, (2) Environmental impact category weight is BEES Stakeholder Panel, (3) Discount rate is 3.0%, (4) Major group element is Substructure, (5) Group element is Foundations, (6) Individual element is Slab on Grade, (7) Generic 100 % Portland Cement, Generic 5% Limestone Cement, Anonymous IP Cement Product, and (8) Transportation distance from manufacture to use is 50 miles in Figure 4-24 through Figure 4-27. The other two transportation distances are also 50 miles. In Figure 4-28, three summary graphs are chosen and then the results can be shown in Figure 4-29 through 31.

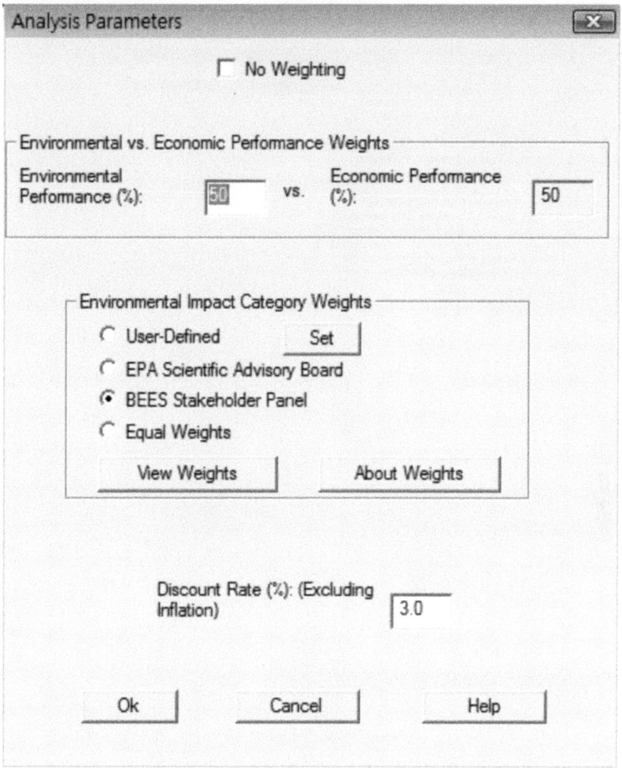

Figure 4-24. Analysis Parameters

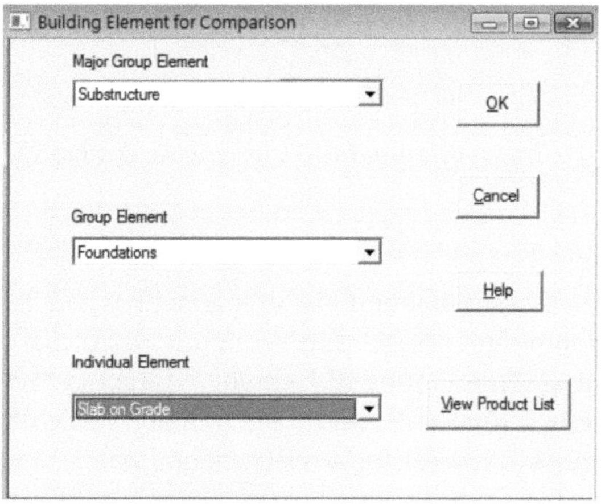

Figure 4-25. Building Element for Comparison

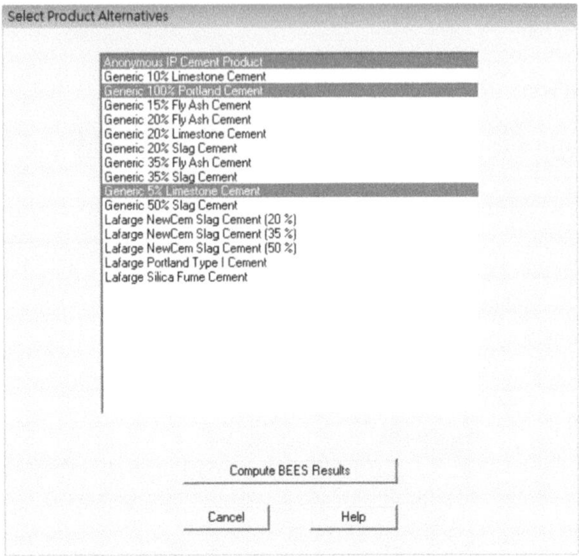

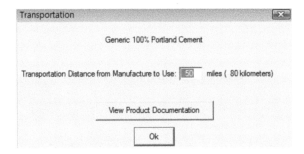

Figure 4-26. Select Product Alternatives

Figure 4-27. Transportation for Generic 100% Portland Cement

The product has a better LCA assessment when it has a lower value for environmental impact and economic performance. Thus, the summary value of Anonymous IP Cement Product is lowest for overall performance in Figure 4-29, and the summary value of Anonymous IP Cement Product is lowest for environmental performance in Figure 4-30. The values for economic performance are the same in Figure 4-31, so, for example here 1.84 is the lowest cost.

Figure 4-28. Select BEES Reports

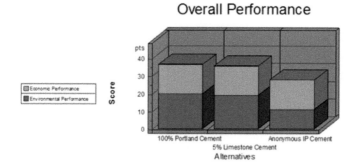

Note: Lower values are better

Category	100% OPC	5% Lime	Anon IP
Economic Perform.--50%	16.7	16.7	16.7
Environ. Perform.--50 %	19.9	19.0	11.1
Sum	36.6	35.7	27.8

Figure 4-29. The Values for Overall Performance

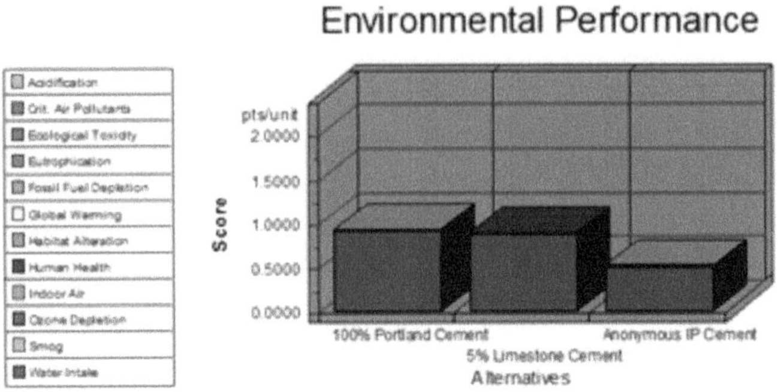

Environmental Performance

Note: Lower values are better

Category	100% OPC	5% Lime	Anon IP
Acidification--3%	0.0000	0.0000	0.0000
Crit. Air Pollutants--9%	0.0006	0.0006	0.0005
Ecolog. Toxicity--7%	0.0025	0.0024	0.0019
Eutrophication--6%	0.0004	0.0004	0.0004
Fossil Fuel Depl.--10%	0.0008	0.0008	0.0007
Global Warming--29%	0.0049	0.0047	0.0038
Habitat Alteration--6%	0.0000	0.0000	0.0000

Press PageDown for more results...

Human Health--13%	0.9165	0.8714	0.5078
Indoor Air--3%	0.0000	0.0000	0.0000
Ozone Depletion--2%	0.0000	0.0000	0.0000
Smog--4%	0.0006	0.0006	0.0005
Water Intake--8%	0.0001	0.0001	0.0001
Sum	0.9264	0.8810	0.5157

Figure 4-30. The Values for Environmental Performance-cont.

ATHENA Impact Estimator for Buildings (ATHENA)

ATHENA is a decision support tool that assists with decisions about the selection of material mixes and provides a cradle-to-grave process for a whole building over its expected life. It is a life cycle tool in green building to accomplish two objectives: life cycle cost (LCC) and life cycle environmental impact (LCEI). A user can evaluate various design options to decrease LCEI. It provides inputs for different materials and design options, and it allows users to change designs, use different materials and make side-by-side comparisons. The environmental impacts consider material manufacturing, transportation, on-site construction, maintenance, repair and disposal.[23]

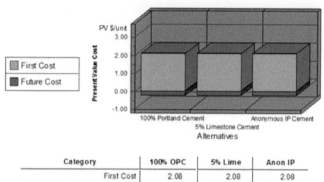

Category	100% OPC	5% Lime	Anon IP
First Cost	2.08	2.08	2.08
Future Cost-- 3.0%	-0.24	-0.24	-0.24
Sum	1.84	1.84	1.84

Figure 4-31. The Values for Economic Performance

An example is from the ATHENA version 4 tutorial. The ATHENA main screen includes five primary items, such as project name, project location, floor area, building life and building type. Although ATHENA does not include an operating energy simulation capability, a building operating energy consumption screen in Figure 4-32 can be used to compute the fuels. In Figure 4-33, the "Add Assembly" menu shows six important assemblies, foundations, walls, mixed columns and beams, roofs, floors, and extra basic materials. Besides, two wood stud walls can also be added. For example, exterior wood stud wall screen is shown in Figure 4-34, and interior wood stud wall screen is shown in Figure 4-35. Figure 4-36 illustrates wood stud is shown in "Available Assembly Components" list, and "Used Components" list shows exterior and interior wood stud walls. Moreover, Figure 4-36 shows ATHENA can combine many different wall sub-assemblies into a custom wall assembly. Other

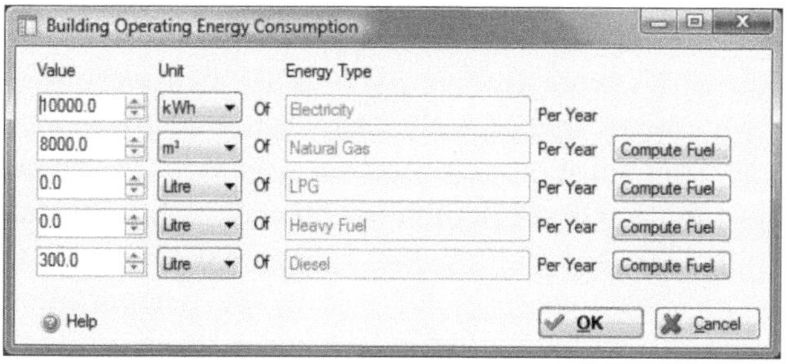

Figure 4-32. Building Operation Energy Consumption

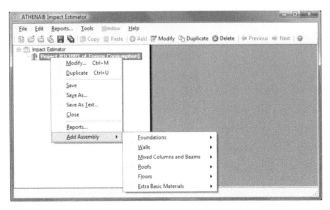

Figure 4-33. Add Assembly

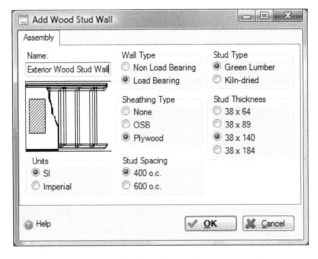

Figure 4-34. Add Wood Stud Wall-Exterior

detailed setup information is shown from ATHENA tutorial.

In Figure 4-37, there are five parts to set the report. In particular, "Summary Measures" includes eight environmental impact measures, such as energy consumption, acidification potential, global warming potential, HH response effects potential, ozone depletion potential, smog potential, eutrophication potential and weighted resource use. Figures 38 and 39 show three items which are material, transportation, and annual operating energy appearing in five life cycle stages with two different measurement charts. In Figure 4-40, the report shows the quanti-

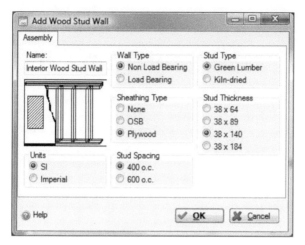

Figure 4-35. Add Wood Stud Wall-Interior

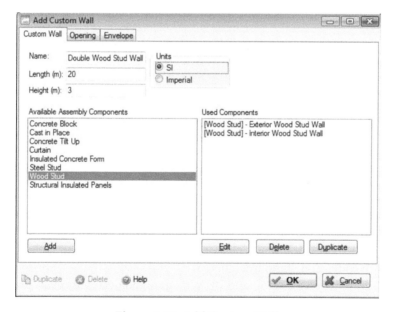

Figure 4-36. Add Custom Wall

ties of materials which are used from project 1. ATHENA also provides comparison reports. The reports and graphs are in Figures 41 through 43. There are two projects in the tutorial. Project 2 has low manufacturing in primary energy consumption and global warming potential.

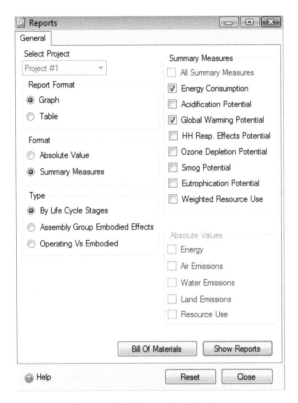

Figure 4-37. Reports Window

CONCLUSIONS

　　Building energy analysis and LCA platforms have increased significantly in scope and complexity since 2000 and have benefited from the confluence of standards and performance measurement techniques. As with codes and standards, there is no one system that authoritatively defines the whole science, and BEA, despite major advances in graphical interfaces and integration with design tools, remains a challenging and complex endeavor. One of the next developments may be to integrate BEA and LCA platforms with compliance options for ASHRAE 189.1 and the IgCC. That way, designers and energy engineers could conduct performance evaluations which automatically adjust for compliance with the baseline codes and standards. This would streamline cost estimating for various options.

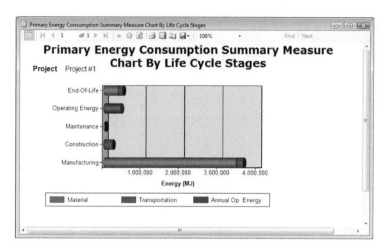

Figure 4-38. Primary Energy Consumption Summary

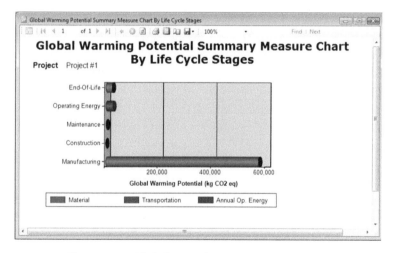

Figure 4-39. Global Warming Potential Summary

References

1. http://apps1.eere.energy.gov/buildings/tools_directory/alpha_list.cfm
2. *Procedures for Commercial Building Energy Audits*, 2nd Edition, American Society of Heating Refrigerating and Air-Conditioning Engineers, 2011.
3. Krarti, M., *Energy Audit of Building Systems: An Engineering Approach*. CRC Press, 2000.
4. http://www.ashrae.org/publications/page/pcbeabook
5. "Hourly simulation" refers to a function of BEA which provides hourly results of energy consumption and performance.

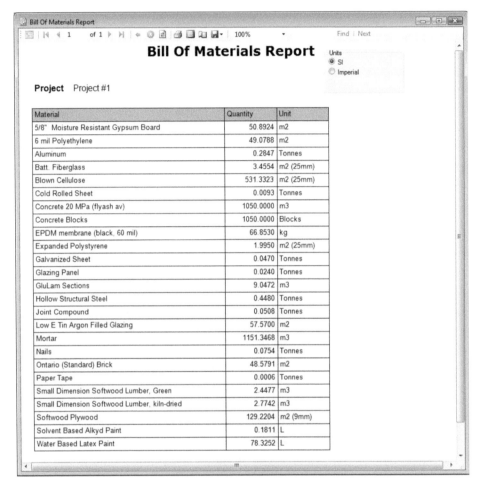

Figure 4-40. Bill of Materials Report

6. *Procedures for Commercial Building Energy Audits*, 2nd Edition, American Society of Heating Refrigerating and Air-Conditioning Engineers, 2011.

7. *Ibid.*

8. © 2003 Washington State University Cooperative Extension Energy Program. This material was written and produced for public distribution. You may reprint this written material, provided you do not use it to endorse a commercial product. Please reference by title and credit Washington State University Cooperative Extension Energy Program. Published May 2003. WSUCEEP2003-049

9. Acknowledgement: Rocky Mountain Institute, Integral Group, Taylor Engineering, kW Engineering

10. http://apps1.eere.energy.gov/buildings/tools_directory/software.cfm/ID=58/ www.imagicdesign.co.jp/trnsys/

11. http://sel.me.wisc.edu/trnsys/features/

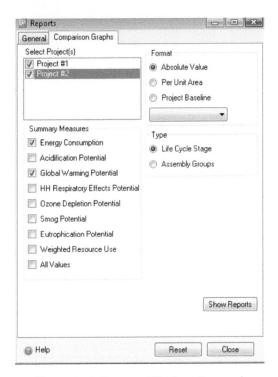

Figure 4-41. Reports Window-Comparison

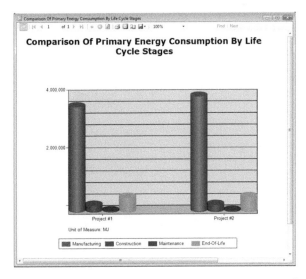

Figure 4-42. Comparison of Primary Energy Consumption

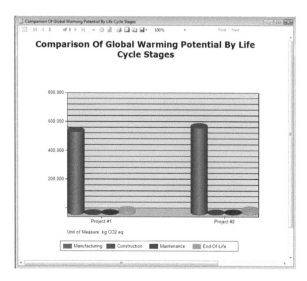

Figure 4-43. Comparison of Global Warming Potential

12. The NFRC is a non-profit organization that administers a uniform, independent rating and labeling system for the energy performance of windows, doors, skylights, and attachment products including shading systems. www.nfrc.org. NFRC has developed the Component Modeling Approach Software Tool (CMAST), which establishes a set of performance libraries of approved components (frames, glass, and spacer) which can be accessed for configuring fenestration products for a project, and obtaining a U-factor, Solar Heat Gain Coefficient (SHGC), and Visible Transmittance (VT) rating for those products, which can then be reflected in a CMA Label Certificate for code compliance. CMAST is web-based as well as client-based. www.nfrc.org/CMAprogram.aspx

13. See: http://www.iso.org/iso/iso_catalogue/catalogue_tc/catalogue_detail.htm?csnumber=26425

14. SEE: http://windows.lbl.gov/software/window/window.html

15. THERM6.3/WINDOW6.3 NFRC Simulation Manual January 2011, Lawrence Berkeley National Laboratory.

16. http://gundog.lbl.gov/dirsoft/d2whatis.html#What Is DOE-2?

17. http://apps1.eere.energy.gov/buildings/energyplus/energyplus_about.cfm.

18. http://sketchup.google.com/green/analysis.html

19. http://usa.autodesk.com/adsk/servlet/pc/index?id=11179508&siteID=123112

20. http://apps1.eere.energy.gov/buildings/energyplus/energyplus_about.cfm

21. http://www.lisa.au.com/

22. http://www.nist.gov/el/economics/BEESSoftware.cfm/

23. http://www.athenasmi.org/our-software-data/impact-estimator/

Chapter 5

Commissioning

Jason Thompson, P.E.

INTRODUCTION

All buildings go through a life cycle—from initial construction through years of operation—which may include multiple occupancy requirements and renovations. To ensure that buildings are operated as efficiently as possible, the energy consuming systems should be evaluated and "fine-tuned" at various times throughout the life of a building. As an old proverb says, "If we do not change our direction, we are likely to end up where we are headed." Although this sounds like basic common sense, this is very true of commissioning in that the issues that affect building performance will only continue, or even get worse over time, unless systems are properly commissioned and maintained. If nothing is done to change the trend or direction, building owners and operators should not be surprised when the same problems and high operating costs recur year after year.

One can also think about commissioning in terms of owning and maintaining a vehicle. When purchasing a car, especially a previously owned model, many dealerships highlight that the vehicle is "pre-certified." This means that it has been through a multi-point inspection, and everything has been checked out to ensure that it operates properly. As the mileage adds up, however, systems may begin to deteriorate, and the vehicle needs a tune-up from time to time to ensure that it continues to operate at a high performance level and provide the best fuel economy. Similarly, buildings also need to undergo an up-front commissioning process and periodic "re-commissioning" to optimize performance and reduce energy consumption.

In 2009, a study was conducted for the California Energy Commission by Lawrence Berkeley National Laboratory to address the results of building commissioning.[1] This study covered 643 buildings, 99 million square feet, and 26 states. After reviewing the findings, one of the main conclusions from this study was that "commissioning is arguably the single-most cost-effective strategy for reducing energy, costs, and greenhouse gas emissions in buildings today." This conclusion emphasizes the importance of commissioning as a tool to optimize the performance of both new and existing buildings.

In order to optimize building performance, a systematic commissioning process should occur at the time of initial building construction to make sure the systems are installed correctly and to verify that everything functions according to the original design intent. Buildings are dynamic, however, and the building requirements can change over time. Building systems, even if originally operated properly, tend to degrade over time. As such, the building systems should be periodically evaluated to ensure they meet the current requirements of the facility. This provides the basis for building commissioning, retro-commissioning, and ongoing commissioning, and this chapter is dedicated to the important role commissioning plays throughout the life cycle of a building.

TYPES OF COMMISSIONING

Before going into too much detail, it's important to have some good working definitions of various types of commissioning. There may be slight variations depending on the source consulted, but following are the definitions used for this chapter to provide a consistent frame of reference moving forward:

Commissioning—A quality-focused process for verifying and documenting that the facility and its systems are planned, designed, installed, operated and maintained to meet the owner's requirements.

Retro-commissioning (RCx)—Commissioning applied to an existing facility, whether previously commissioned or not, to help the facility and its systems meet the owner's current and anticipated future requirements (not necessarily its original design).

Ongoing Commissioning—A continuation of commissioning into the occupation and operation phase, or a continuation of retro-commissioning, to verify that the facility and its systems continue to meet the owner's current and evolving requirements.

Based on the definitions above, commissioning takes place when systems are initially installed. This could include new building construction or the installation of new equipment. Retro-commissioning applies to existing buildings, and these buildings may or may not have been previously commissioned. If a building has been previously commissioned, the term "re-commissioning" is sometimes used in the industry, but since the working definition of retro-commissioning also incorporates this option, re-commissioning is not included as a separate item of discussion in this chapter.

The matrix in Table 5-1 summarizes the different commissioning options that are available for various facilities.

Table 5-1. Matrix of Commissioning Options

Type	New Construction	Existing Building	Previously Commissioned	Not Previously Commissioned	Commission to Owners Original Requirements	Commission to Owner's Current Needs
Commissioning	✓				✓	
Retro-commissioning		✓	✓	✓	✓	✓
Ongoing Commissioning		✓	✓		✓	✓

PART 1: COMMISSIONING

Why Commissioning?

Commissioning for new buildings involves review and documentation of the design and construction of a building, with a focus on ensuring that the building's performance meets the requirements of the building owner and operator. A focus is also placed on the energy efficient operation of equipment and maintaining a safe work environment. It is important for a building to be properly commissioned for several reasons:

a) Commissioning helps ensure that the equipment will perform with the original intention at the start of the building's life cycle. The more "out of tune" the building, the quicker the building systems will deteriorate causing operational and comfort issues that can worsen over time.

b) Commissioning can help ensure that energy efficient designs are maintained throughout the construction process. It's not uncommon to find that energy efficiency design measures were not implemented during the construction phase due to cost overruns. This "value engineering" process might save initial costs, but the building owner may pay for this with increased utility costs, maintenance expenses, and potential comfort issues, often for many years, down the road.

c)   New building control systems are becoming more and more sophisticated, and these systems require expertise to ensure they are set up correctly and operating properly prior to occupancy of the building.

Commissioning is therefore vital to the immediate and long-term performance of a new building. Although many owners may ask if they can afford to go through the commissioning process, a better question may be, "Can we afford *not* to properly commission this building based on the future costs and operational issues that will likely result otherwise?"

What is the Process?
Commissioning is a process that begins with the design process and continues through the construction phase and operation as a functional building. An outline of the commissioning process for new buildings is provided in Figure 5-1.

What are the Benefits?
Everyone typically benefits from commissioning of a new building—maybe with the exception of contractors that regularly return to perform ongoing repairs on equipment. Building owners benefit through lower utility and maintenance costs, building operators benefit from reduced time spent trying to resolve problem areas, and the building occupants benefit from better indoor air quality. Although not a comprehensive list, Table 5-2 is a summary of some of the major benefits of commissioning.

Savings and Costs
Since savings from commissioning are difficult to predict, savings and costs are better expressed in terms of a range of values. Table 5-3 for the California Commissioning Collaborative provides some typical savings based on a comprehensive study.

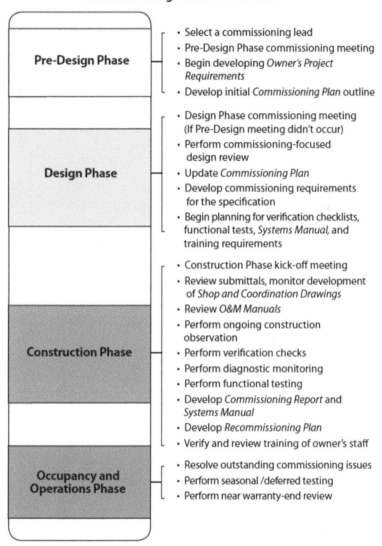

Commissioning Process Overview

Pre-Design Phase
- Select a commissioning lead
- Pre-Design Phase commissioning meeting
- Begin developing *Owner's Project Requirements*
- Develop initial *Commissioning Plan* outline

Design Phase
- Design Phase commissioning meeting (If Pre-Design meeting didn't occur)
- Perform commissioning-focused design review
- Update *Commissioning Plan*
- Develop commissioning requirements for the specification
- Begin planning for verification checklists, functional tests, *Systems Manual,* and training requirements

Construction Phase
- Construction Phase kick-off meeting
- Review submittals, monitor development of *Shop and Coordination Drawings*
- Review *O&M Manuals*
- Perform ongoing construction observation
- Perform verification checks
- Perform diagnostic monitoring
- Perform functional testing
- Develop *Commissioning Report* and *Systems Manual*
- Develop *Recommissioning Plan*
- Verify and review training of owner's staff

Occupancy and Operations Phase
- Resolve outstanding commissioning issues
- Perform seasonal /deferred testing
- Perform near warranty-end review

Figure 5-1. Overview of Commissioning Process[2]

Table 5-2. Benefits of Commissioning

1. Lower energy usage and reduced utility bills
2. Fewer operating issues and problem areas with building systems
3. Increased occupant or tenant satisfaction
4. Better working conditions and indoor air quality

Table 5-3. Typical Commissioning Savings and Costs[3]

Description	Range of Values
Value of Energy Savings	$0.02 to $0.19 per sqft
Value of Non-Energy Savings	$0.23 to $6.96 per sqft
Total Commissioning Costs	$0.49 to $1.66 per sqft

It should be noted that these savings and costs may vary in different parts of the country due to diverse energy rates and labor costs. The largest part of the commissioning cost involves the service provider fee, which can be 75% or higher of the total cost.

In a 2009 Lawrence Berkeley National Laboratory study, the average results for commissioning of new buildings were compiled over the large database of buildings that was part of this study. See Table 5-4.

Table 5-4. Average Commissioning Results from LBNL Study[4]

Median costs	$1.16/square foot
Median energy savings	13%
Median payback time	4.2 years

Getting Started

Once a decision to participate in the commissioning process has been made, there are several ways that the owner can be involved to ensure a successful project. Table 5-5 is a summary of some of the major steps that can be taken.

PART 2: RETRO-COMMISSIONING (RCX)

The Case for RCx

Retro-commissioning for existing facilities helps the facility and its systems operate at an optimum level to meet the owner's current and

Table 5-5. Key Steps to Get Started

1. Determine if local, utility, or government incentives are available to offset the costs;
2. Commit resources well in advance of project planning/design;
3. Contract with a qualified commissioning provider;
4. Ensure that the commissioning scope is in line with the owner's objectives;
5. Commit in-house personnel to support the commissioning provider and obtain buy-in from the organization;
6. Ensure that the commissioning requirements are included in contractor specifications.

anticipated future requirements (not necessarily its original design). As with commissioning of new buildings, priority is placed on energy efficient operation of equipment and maintaining a safe work environment.

As buildings get older, they may develop symptoms that indicate that things are not working properly from both energy and operational standpoints. Just as doctors diagnose potential health problems, "building doctors" may also be needed to assess and correct the issues negatively affecting the performance of existing buildings. The RCx process can therefore be likened to a "check-up" that can identify pre-existing conditions and prevent even more problems that may arise in the future.

Retro-commissioning is a key part of optimizing the overall performance of an existing building for the following reasons:

*   Retro-commissioning offers property owners and building operators a way to improve building efficiency, comfort, and reliability at minimal implementation costs;

*   And the best part—these benefits go hand-in-hand with reduced energy consumption and lower operating costs.

RCx Process

The retro-commissioning process typically involves several distinct phases. These may be slightly different based on the RCx service provider or utility company requirements if conducted as part of a utility-sponsored program, but the basic concept still applies. The four phases include:

1)   Planning and initial screening;
2)   Site assessment and investigation;
3)   Implementation of recommended measures; and
4)   Measurement and verification of results and final reporting.

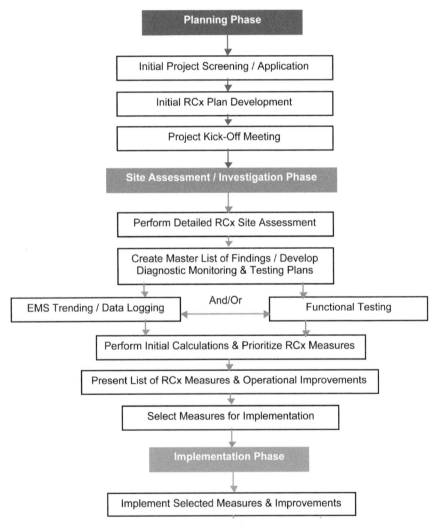

Figure 5-2. Retro-commissioning Process Overview (*Continued*)

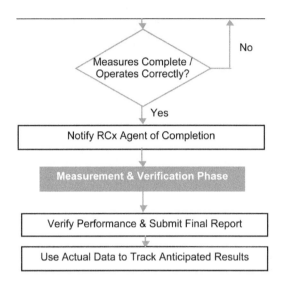

Figure 5-2. Retro-commissioning Process Overview (*Concluded*)

Planning/Screening Phase

The planning and application phase involves screening a potential building as a good candidate for an RCx study by reviewing utility bills, documenting preliminary survey information, and discussing operations with facility personnel. If the facility is applying as part of a utility-sponsored RCx program, an application is usually submitted and reviewed as part of the qualification process. Preliminary scope meetings may take place. A preliminary RCx plan, including projected energy savings and RCx costs, may also be generated as part of this phase. A preliminary site visit may be conducted to assess the initial opportunities and outline the scope and project objectives.

Assessment Phase

The assessment phase involves a detailed on-site survey by a qualified RCx service provider. The service provider inspects the operation of the current HVAC equipment and determines if there are operating deficiencies or opportunities for energy conservation with a focus on low-cost measures. Simple no-cost repairs may be completed during this assessment. The sample data collection form for an air-handling unit (Figure 5-3) illustrates the level of detail that may be collected from a typical unit during a site visit.

AIR HANDLING UNIT INFORMATION FORM

Facility: _____ **Date:** _____

Building: _____

AIR HANDLING UNIT **Unit Tag:** _____

　　MFGR _____

　　MODEL NBR _____ Area Served _____

1 System Type:					DESIGN	ACTUAL / MEASURED
☐ Constant Volume	☐ Variable Air Volume		Total Air flow (CFM)			
☐ Changeover Bypass	☐ Multi-zone	☐ Dual Duct	O.A. Air flow (CFM)			
2 Cooling Coil Type:			TSP (in.)			
☐ Direct Expansion (DX)	☐ Chilled Water	☐ None	Fan BHP / Amps			
3 Heating Coil Type:			Motor HP			
☐ Hot Water ☐ Electric	☐ Steam	☐ None	Motor RPM / VFD Output (Hz)			
4 Other Coils:			Motor Voltage			
☐ Heat Recovery	☐ None		Motor Efficiency			
5 Economizer Type:			Cooling Capacity (MBH)			
☐ Dry Bulb	☐ Enthalpy	☐ None	Cooling Coil EDB (T1)			
6 Economizer Setpoint			Cooling Coil LDB (T2)			
°F or BTU/Lb			CHW EWT			
7 Supply Air Temp (VAV, MZ, DD)			CHW LWT			
Setpoint ____ °F Actual / Measured ____ °F			CHW Flow (GPM)			
Space Temp Setpoint (CV)			CHW Press Drop (ft)			
Setpoint ?? °F Actual / Measured ____ °F			DX Coil Suction Temp			
8 Fan Volume Control Method			Htg Coil Capacity (MBH)			
☐ VFD ☐ IGV	☐ Other	☐ None	Heating Coil EAT			
9 Fan Volume Control Parameter			Heating Coil LAT			
☐ Supply duct static pressure	☐ Other	☐ None	HW EWT			
10 Control Parameter			HW Flow (GPM)			
Setpoint ____ in. w.g. Actual / Measured ____ in. w.g.			HW Delta-T			
11 CHW Valve Type			Htg Coil Press Drop (ft)			
☐ 2-way ☐ 3-way	☐ None		Electric Coil Capacity (kW)			
12 CHW Valve Control Type			Electric Coil Stages			
☐ Pneumatic ☐ Electric/Electronic	☐ DDC		Position			

13 Filter Type				
☐ Throwaway - fiberglass ☐ Throwaway - pleated media 2" ☐ Throwaway - pleated media 4" ☐ Other	Filter Condition: ☐ Clean ☐ Dirty ☐ Very dirty	Schedule:	ON ____ AM / PM	OFF ____ AM / PM

NOTE:

Figure 5-3. Sample Data Take-off Form for RCx Survey (*Continued*)

SINGLE ZONE AIR HANDLING UNIT

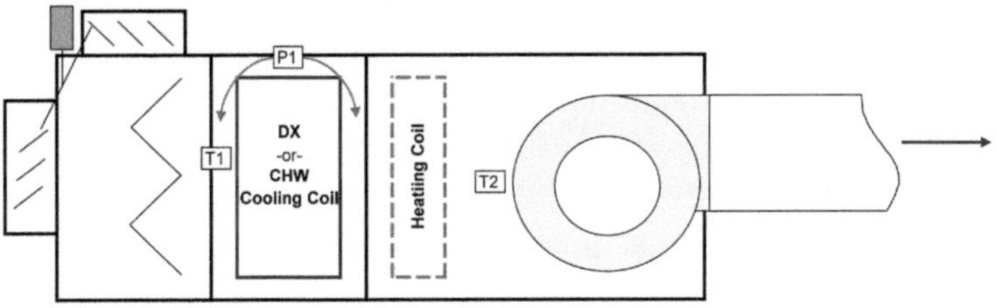

P1

DX
-or-
CHW
Cooling Coil

T1

Heating Coil

T2

Actuator Type:	☐ Pneumatic ☐ Electric ☐ None			
Damper/Actuator Condition:	Actuator Operative	☐ YES ☐ NO	Economizer switchover at setpoint	☐ YES ☐ NO
	Dampers Operative	☐ YES ☐ NO		
Controls Condition:	Controls Functional	☐ YES ☐ NO	CHW Valve modulates thru full range	☐ YES ☐ NO
Unit Condition:	Drain pan drains properly	☐ YES ☐ NO	Access doors close tightly	☐ YES ☐ NO
	Drain pan rusting out	☐ YES ☐ NO	Significant duct leakage	☐ YES ☐ NO
Clg Coil Condition:	Excessively dirty	☐ YES ☐ NO	Fan/Motor Alignment OK	☐ YES ☐ NO
	Fins deteriorating	☐ YES ☐ NO		
	Sight glass clear (DX)	☐ YES ☐ NO	Safety Labels Affixed	☐ YES ☐ NO
Fan/Motor/Belts Condition:	Bearings noisy	☐ YES ☐ NO	Space Temp	Actual Space
	Belt(s) loose	☐ YES ☐ NO	Setpoint _____	Temperature _____

MULTI-ZONE AIR HANDLING UNIT

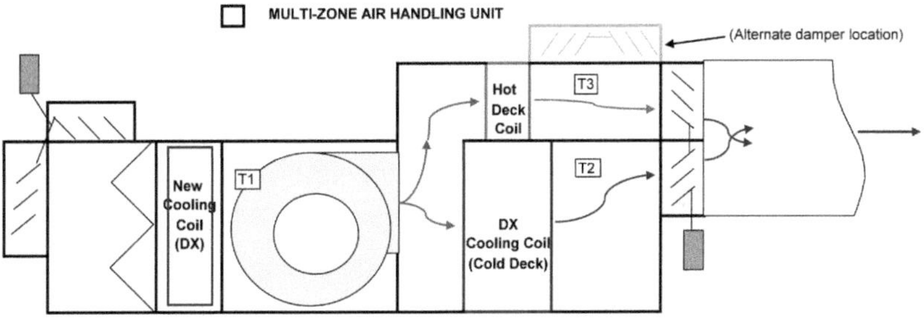

(Alternate damper location)

Hot
Deck
Coil

T3

T1

New
Cooling
Coil
(DX)

DX
Cooling Coil
(Cold Deck)

T2

		Space Temp Setpoints	Actual Space Temperature		Space Temp Setpoints	Actual Space Temperature
Number of zones: _____ (multi-zone actuators)		Zone 1 _____ _____		Zone 5	_____ _____	
		Zone 2 _____ _____		Zone 6	_____ _____	
Hot Deck Energized: ☐ YES ☐ NO		Zone 3 _____ _____		Zone 7	_____ _____	
		Zone 4 _____ _____				

Take-off Notes / RCx Opportunities:

Non-functioning Items:

Figure 5-3. Sample Data Take-off Form for RCx Survey (*Concluded*)

During the investigation, the RCx service provider performs functional testing of HVAC equipment, evaluates the performance as it relates to original design, and evaluates the condition of the current control systems. The investigation also involves analyzing trend data from the building automation system (BAS) and interacting with the controls vendor to determine how the systems currently operate. Trending of equipment operation may also be used during this phase to develop a baseline prior to implementation of recommended measures.

The intent of the assessment is to determine the magnitude of the potential energy savings and to develop a list of retro-commissioning measures (RCMs). The final deliverable is a report that summarizes the current equipment condition and presents the recommended RCMs along with energy savings potential (often called the master list of findings). The assessment provides the basis for the tasks that will be completed in the subsequent phases.

Implementation Phase

After the site investigation, the next phase involves implementation of the various RCMs. As part of this process, a follow-up visit by the RCx service provider may be required to further develop the RCMs that have the best energy savings potential and provide more detailed scopes of work. Although several of the measures may be implemented in-house, this phase will likely require the involvement of a contractor.

Measurement and Verification Phase

The final phase of the RCx process involves measurement and verification (M&V) of the RCMs that were implemented. Trending of equipment operation is typically conducted to compare with the pre-implementation trends. Measurements are also taken, which may include amperage, voltage, and power readings. The deliverable is a final M&V report that documents the implementation of the measures and the verified savings.

Typical RCx Measures

Table 5-6 provides a list of some of the most common measures that may be identified during the RCx process. Although certainly not a comprehensive list, this shows the low-cost and no-cost nature of many of the recommendations.

Table 5-6. Top 10 Common RCx Measures

1. Schedule HVAC Systems
2. Adjust Outside Air to Current Needs/Right-sizing Equipment
3. Mitigate Simultaneous Heating and Cooling
4. Reduce Flow from Oversized Pumps
5. Reset Supply Air Temperatures
6. Enable/Optimize Economizer Controls
7. Reset Static Pressures/Optimize VFD Operation
8. Stage Chillers Properly
9. Lower Condenser Water Setpoints
10. Correct Lighting Control Operation

Trending and Diagnosis

As mentioned previously, trending is a vital component of the RCx process. Most large buildings with sophisticated HVAC equipment have a central building automation system (BAS) that is used to monitor and control the major equipment. The front-end BAS is typically located on a computer in the engineering area. Whether the equipment controls are digital, electric, pneumatic, or a hybrid system, the features of the BAS typically include start/stop control, time-of-day scheduling capability, and monitoring and control of chiller plants, boiler plants, and air-handling systems. The BAS also typically provides for the adjustment of temperature and control setpoints for air-handling units and terminal units.

The software program in most modern building automation systems allows for trending of select control points, such as temperatures, operating status, amperage or kW (if input to BAS), VFD percent speed, damper and valve percent open, etc. Sometimes trends can be difficult to set up properly, so the controls vendor should be consulted to ensure that the appropriate trends are set up. Trending can be used in the RCx process in two main ways:

1) Trending can be used to help identify RCx measures by revealing operating deficiencies;

2) Trending can allow for pre- and post-implementation comparison of various operating points to determine the actual savings in the M&V process.

Figures 5-4 and 5-5, which are derived from actual trend data collected during RCx site visits, provide a snapshot of how trending can be used in the RCx process to support RCx recommendations and the M&V process

Figure 5-4 shows variable frequency drive (VFD) fan speed (blue trend line, dark to the right) overlaid with supply air temperature (orange trend line, dark to the left) developed from actual control point data collected during an RCx site visit. The graph shows that the fan operates on a pre-defined schedule from 7 am to 6 pm on a daily basis, as the fan speed drops to zero every day during this period and the discharge air temperature remains at a relatively constant room temperature. The trend data illustrate that the supply air temperature varies from approximately 60°F to 67°F during fan operating periods. The VFD percentage remains fixed at 90%, however.

Since this is a variable air volume system, the fan speed should ramp up or down in response to load. Since the fan speed remains close to 100%, this indicates a problem in the current system configuration or control of the fan speed. After investigation, it was determined that the static pressure sensor was located at the discharge of the fan, and the control was not functioning properly. This shows how trending can be used during the RCx process to help identify operating deficiencies and develop retro-commissioning measures for implementation. After implementation of this RCx measure, the trends should show the VFD speed fluctuating in response to the building load throughout the occupied schedule.

Figure 5-5 shows the chiller amperage draw for operating Chiller 3 (small air-cooled chiller) as a base load chiller with one additional centrifugal chiller to meet the remaining load, versus operating only one centrifugal chiller during normal building operating hours. The building originally had only two large centrifugal chillers, and a small air-cooled chiller was installed to meet the after-hours cooling load for the computer room air-conditioning, units that operate 24 hours per day. Although a good concept for nighttime operation, the sequencing was set up to allow the small air-cooled chiller to operate as the base load chiller 24 hours per day. The air-cooled chiller is less efficient than the water-cooled centrifugal chiller during daytime hours, and Figure 5-5 shows that the overall chiller electric power draw was reduced by over 100 amps when the chiller sequencing was modified to allow the small chiller to only operate during the evening hours from 6 pm to 6 am. This

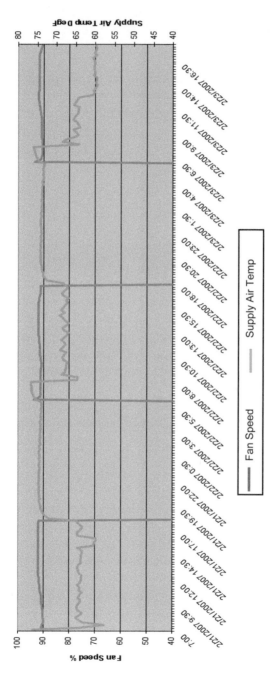

Figure 5-4. Trending of Supply Fan Speed and Supply Air Temperature

example shows how trending can be used during the M&V process to support the verified savings.

What are the Benefits?

There are many obvious benefits to retro-commissioning (abbreviated as RCx), but this section focuses on how RCx can solve the following three common building problems as follows:

1)   Energy Waste

2)   Operating Deficiencies

3)   Comfort Issues

Energy Waste

Building Energy Usage

Energy usage and costs are a major concern with owning and operating buildings. The mechanical systems and energy usage can vary widely between similar types of buildings, which is evidenced in Figure 5-6.

Figure 5-6 illustrates the large range of energy usage between commercial buildings. The scale at the bottom shows both the percentile ranking and energy utilization index (EUI) in kBtu/sqft/yr. The buildings in the shaded section to the left represent the top performing buildings that are in the top 25th percentile in terms of energy usage. Conversely, the buildings in the shaded section to the right represent the buildings that are in the bottom 25th percentile in terms of energy usage.

There are a couple of important conclusions that can be drawn from these data. First, there are a significant number of buildings that are below the top 25th percentile that have a lot of room for improvement. Secondly, the gap is immense between the top performing buildings and the bottom performing buildings. The energy usage ranges from approximately 30 kBtu/sqft/yr to approximately 340 kBtu/sqft/yr, indicating that many of the lower performing buildings are performing extremely poorly. This energy distribution makes a good case for retro-commissioning of buildings, as up to 15% savings, or often even more, can be achieved through implementation of low-cost and no-cost RCx measures.

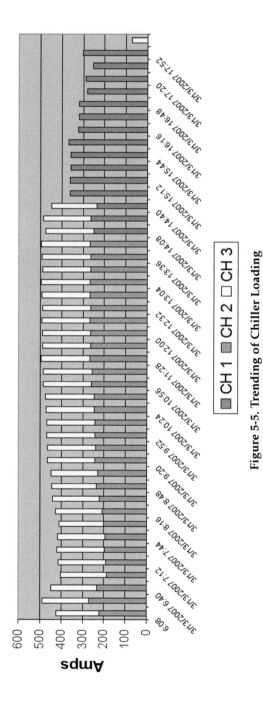

Figure 5-5. Trending of Chiller Loading

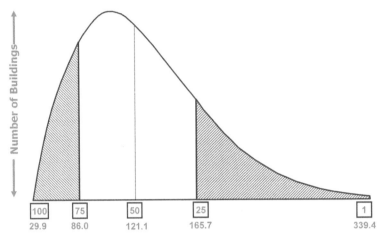

Figure 5-6. Commercial Building Energy Intensity[5]

Technology vs. Performance

Many building owners have the mindset that there may not be much opportunity for energy savings because the facility is either new or has already installed energy-efficient upgrades and controls to building systems. Although new equipment and controls have inherent advantages, age and equipment are not necessarily significant predictors of how much energy a building uses. Many buildings with brand new control systems, VFDs on motors, etc. may use as much or more energy than buildings with older systems. This observation is evidenced in Figure 5-7.

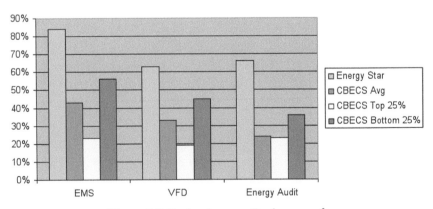

Figure 5-7. Technology vs. Performance[6]

The green bars show ENERGY STAR® certified buildings. As expected, almost 80% of these buildings have energy management systems (EMS), and over 60% of these buildings have VFDs on equipment and have had an energy audit. Although it may seem counterintuitive, over 50% of the bottom 25% of buildings also have an EMS and over 40% have VFDs. Only a small percentage of the top performing 25% of buildings have upgraded technology. These data illustrate that there is still of lot of opportunity to fine-tune systems and increase the energy performance, even in newer buildings or facilities with upgraded technology.

Operating Deficiencies & Occupant Comfort Issues

In addition to energy waste, operating deficiencies and occupant comfort issues present difficulties for buildings that can be addressed with RCx. Many buildings were never commissioned properly when originally constructed, which leads to ongoing problems with meeting heating or cooling loads. Another reason to consider RCx is that the owner's operating requirements may have changed over time. It was mentioned earlier that buildings are dynamic. It may be that occupancy conditions have changed since the building was originally constructed. Space types may have also changed—i.e., offices may have been converted into conference rooms, etc. Design standards have also changed over the years, so the standards that were applicable when the facility was originally built may not be the same today. RCx can also address occupant comfort issues that are associated with many of these operating deficiencies noted above. In many cases, tenant hot and cold calls can be reduced by going through the RCx process, and overall comfort conditions can be improved.

RCx Economics

Retro-commissioning typically focuses on low-cost and no-cost measures, so the implementation costs are typically low and the payback period should be attractive from a financial perspective. In a 2009 Lawrence Berkeley National Laboratory study, the average results for retro-commissioning of existing buildings were compiled over the large database of buildings that was part of this study. See Table 5-7.

In 2009, a study was conducted by the Texas A&M Energy Systems Laboratory to summarize the results of retro-commissioning for different building types. The results of this study, summarized in Table 5-8,

Table 5-7. Average RCx Results from LBNL Study[7]

Median costs	$0.30/Square Foot
Median energy savings	16%
Median payback time	1.1 years
Energy Savings persistence	3 to 5 years at least

show that RCx can be effectively applied to a variety of different building types.

Table 5-8. Cost Benefits of RCx from Texas A&M Study[8]

Bldg Type	No. of Bldgs	Savings $/Sq Ft/Yr	Cost $/Sq Ft	Payback Yrs
Hospitals	6	$0.43	$0.47	1.1
Lab/Offices	7	$1.26	$0.37	0.3
Class/Offices	5	$0.43	$0.23	0.5
Offices	8	$0.22	$0.33	1.5
Schools	2	$0.17	$0.34	2.0
Avg/Total	28	$0.54	$0.36	0.7

Summary of Results—RCx Program #1

This section summarizes the results for a utility-sponsored RCx program in California that was conducted with the assistance of an RCx service provider.[9] These results were compiled over a 3-year period and encompassed 15 commercial buildings.

- **Retro-Commissioning Findings**
  - Electricity savings from 1.6% to 9.4%, average 4.6%
  - Natural gas savings from 3.9% to 53.2%, average 10.0%
  - > 3,000,000 kWh saved

- **Retro-Commissioning Costs**
  - Investigation: $0.10/sq ft (paid by utilities)
  - Implementation: $0.12/sq ft

- **Simple Payback Average of 1.5 Years**

Summary of Results—RCx Program #2

This section summarizes the results for another utility sponsored RCx program that was conducted with the assistance of an RCx service provider.[10] These results were compiled over a 2-year period and encompassed three commercial buildings—a high-rise office building, an upscale hotel, and a large data center. It should be noted that all recommended measures were not implemented, but all measures that were implemented went through a detailed measurement and verification (M&V) process to ensure that the reported savings were achieved.

* **Retro-commissioning Findings**
  – Electricity savings average: 5%
  – 1,900,000 kWh saved

* **Retro-commissioning Costs**
  – Investigation: $0.10/sq ft (paid by utilities)
  – Implementation: $0.05/sq ft

Case Study

A multi-tenant office tower participated in a utility-sponsored RCx program from 2009 through 2010. The building was originally constructed in 1979, with a primary use as office space, although there is also retail space and restaurants open for public use. The high-rise office tower has over 800,000 rentable square feet.

The building is an all-electric facility with a central chiller plant and electric baseboard heating along the perimeter of each tenant floor. The base building systems (HVAC and service equipment) consumed approximately 12.7 million kWh in 2009 prior to implementation of any RCx measures. After the building was accepted into the RCx program, the service provider performed an initial retro-commissioning study that revealed two main areas of focus: 1) dedicated nighttime setback schedule for electric baseboard heaters; and 2) optimization of the main air-handling units. In addition, several other low-cost energy saving opportunities were identified, two of which included supply air temperature optimization and chilled water temperature reset.

The AHU optimization measure involved retro-commissioning eight main variable air volume (VAV) air-handling units and their associated return air systems to address suction pressure imbalances and the inability to operate a single set of fans longer and optimally during milder periods of the year. The tasks involved relocating and/or installing new return air duct static pressure sensors; performing test and balance procedures on all fans to determine new static pressure setpoints for the high-rise and low-rise air handling units (AHU); implementing a revised return air fan tracking sequence of operation; providing optimized programming to allow for the increased operation of a single AHU in each low-rise AHU pair; and re-sheaving four air-handling units to set new static pressure setpoints.

Following a review process, the recommended measures were implemented by the facility. A follow-up site visit was then conducted to verify installation and confirm actual energy savings. After implementation of the four RCx measures described above, the facility achieved dramatic energy savings. The base building electric usage for a recent 12-month period (April 2010—March 2011) was 11.6 million kWh compared to the 2009 usage of 12.7 million kWh. This represented a savings of over 1 million kWh per year, which equates to approximately 9% overall energy usage reduction. The implementation cost was just over $60,000 to accomplish the tasks outlined above, which resulted in a payback of only several months.

The graph in Figure 5-8 shows the energy reduction that was achieved for this facility.

Now that the retro-commissioning process is complete, the facility plans to continuously monitor its equipment and systems to maintain energy savings and identify additional energy efficiency improvements.

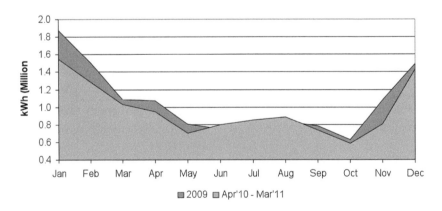

Figure 5-8. Case Study Energy Usage Reduction

PART 3: ONGOING COMMISSIONING

Why is it Important?

One of the questions that frequently comes up during the commissioning or RCx process is "How long will the savings actually last?" That is a very good question because, as previously discussed, buildings are dynamic and performance tends to degrade over time unless steps are taken to ensure the persistence of savings. We intuitively know that setpoints and schedules can be overridden, human interaction can affect systems, and building usage patterns can change. So how can a facility maintain the savings that were achieved after going through a commissioning or RCx process? Ongoing commissioning, or monitoring-based commissioning, can provide a key role in ensuring the ongoing success of a commissioning project.

Persistence of Savings

A study conducted by Texas A & M in 2001 showed that 80% of the savings from RCx measures were still present three to four years later. Similarly, a 2004 study by Sacramento Municipal Utility District (SMUD) indicated that savings began to degrade slightly in the 4th year after implementation. The graph in Figure 5-9 shows the persistence of savings from another case study for a commercial office building. This graph also supports the conclusion that savings may begin to erode, 3 to 4 years after implementation. Note that the graph also shows the tremendous savings that were achieved in the first year immediately

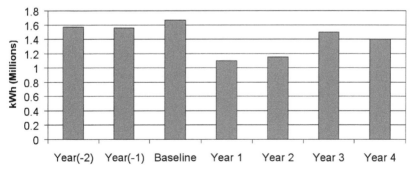

Annual Electricity Use

Figure 5-9. Case Study—Persistence of Savings[11]

following implementation of the RCx measures.

Process & Performance Tracking

Performance tracking is a key element in the ongoing commissioning process. This typically uses computer software to help identify and diagnose problems, and human intervention to determine the steps needed to correct the issues. Software is needed to monitor the performance of specific building systems, and some systems can even automatically notify the building operator if there are any parameters that fall outside of normal expected values. Figure 5-10 is an overview of the performance tracking process.

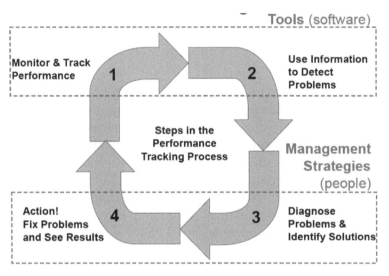

Figure 5-10. Performance Tracking Process[12]

Four strategies that can be used as part of implementing a performance tracking process at a facility are shown in Table 5-9.

Although building engineers would like to better monitor and track energy performance, the only tools typically at their disposal include utility meters and the building automation system, which may not provide very detailed information with regard to the cause of performance issues. The use of predictive diagnostics tools can help sustain the savings and identify further "drift" from normal operation. Table 5-10 gives a summary of benefits that fault detection and diagnosis can provide.

Table 5-9. Performance Tracking Strategies

1. **Energy Benchmarking**—Compare current building energy performance with a previously determined baseline performance, or the performance of similar types of buildings, to determine if there is a major opportunity for improvement.

2. **Evaluation of Energy Data**—Evaluate billing usage history or monitor other utility data to detect patterns of higher than normal energy usage.

3. **Trend Analysis**—Use the building automation system or other data logging devices to record the operating parameters of equipment or systems over time.

4. **Fault Detection and Diagnostics**—Automatically detects system faults and diagnoses cause of the faults.

Table 5-10. Benefits of Fault Detection and Diagnosis

1. Energy savings due to sustaining RCx measures and identification of abnormal operating conditions;

2. Savings due to predictive maintenance versus time-based maintenance;

3. Provides visibility into performance issues and pending failures of capital equipment.

What are the Benefits?

Ongoing commissioning provides a way to ensure that measures implemented through new building commissioning or an RCx process persist over time. Although not a comprehensive list, Table 5-11 gives a summary of some of the major benefits of ongoing commissioning.

Table 5-11. Benefits of Ongoing Commissioning

1. Monitors and tracks performance to ensure that savings from commissioning or RCx are maintained for a longer period of time;

2. Continuously tracks performance to verify savings and identify improvements;

3. Provides greater insight into the operation and performance of the building;

4. Maintains updated building and system documentation;

5. Keeps building staff engaged in the process and helps to ensure they are trained;

6. Determines new opportunities for savings that may arise over time.

Monitoring-based and Ongoing Commissioning

Monitoring-based commissioning (MBCx) is based on actual measurements and diagnostics to help identify issues and additional savings opportunities that may not be identified through a regular commissioning or RCx process. This process uses permanent energy usage evaluation and diagnostic tools for the entire building and equipment. It also uses data from these tools to provide accurate savings for measures that have been implemented and to provide ongoing commissioning. The intent of MBCx is to achieve more persistent energy savings than could be achieved through a manual process alone.

Figure 5-11 from Lawrence Berkeley National Laboratory shows the impact that monitoring-based commissioning (MBCx) and ongoing commissioning can have on sustaining energy reductions and achieving even greater energy savings.

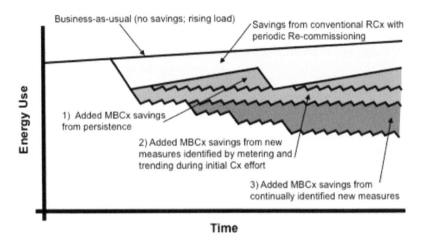

Figure 5-11. Impact of MBCx Over Time[13]

The line at the top represents the "business-as-usual" case, with energy usage gradually increasing over time, since no changes have been implemented. After a conventional commissioning process, a significant amount of savings can typically be achieved. These measures can be expanded by including MBCx activities (including analysis and monitoring of HVAC equipment and building trend data) during the initial commissioning process. By keeping the MBCx efforts in place after the initial study, the recommended measures tend to remain effective over a longer period of time, and new savings may be realized by continually identifying new measures as building usage and occupancy patterns change over time. This maximizes the energy savings potential at the facility over a longer period of time.

CONCLUSION

As outlined in this section, there is a very compelling case for building commissioning. An initial commissioning process should occur at the time of building construction to verify that the systems operate properly. Since systems tend to drift from the original performance over time, periodic retro-commissioning and ongoing commissioning should also be used as tools to ensure that the systems continue to meet the current requirements of the building. By incorporating these processes into a strategic building plan, significant energy savings can be achieved, and most retro-commissioning measures can be implemented for little or no cost. In addition, commissioning can address several of the most common problems facing building owners and operators— energy waste, operating deficiencies, and occupant comfort. Through optimized building performance, building owners and operators can recoup the associated benefits for years to come by monitoring and tracking the results to ensure persistence of savings. In terms of costs versus benefits, commissioning is one of the best things we can do for buildings in order to achieve better energy performance with sustainable results.

References
1. *Building Commissioning—A Golden Opportunity for Reducing Energy Costs and Greenhouse Gas Emissions, Report Prepared for: California Energy Commission Public Interest Energy Research*, Mills, E., Lawrence Berkeley National Laboratory, 2009.
2. *California Commissioning Guide: New Building, Report Prepared for: The California Commissioning Collaborative* Haasl, T., Heinemeier, K., Portland Energy Conservation, Inc., 2006. http://www.documents.dgs.ca.gov/green/commissionguideexisting.pdf
3. *California Commissioning Guide: New Building, Report Prepared for: The California Commissioning Collaborative* Haasl, T., Heinemeier, K., Portland Energy Conservation, Inc., 2006. http://www.documents.dgs.ca.gov/green/commissionguideexisting.pdf
4. *Building Commissioning—A Golden Opportunity for Reducing Energy Costs and Greenhouse Gas Emissions, Report Prepared for: California Energy Commission Public Interest Energy Research*, Mills, E., Lawrence Berkeley National Laboratory, 2009.

http://cx.lbl.gov/documents/2009-assessment/
lbnl-cx-cost-benefit.pdf

5. *Commercial Building Energy Consumption Survey* (CBECS), http://www.eia.gov/
consumption/commercial/

6. *Commercial Building Energy Consumption Survey* (CBECS), http://www.eia.gov/
consumption/commercial/

7. *Building Commissioning—A Golden Opportunity for Reducing Energy Costs and Green-house Gas Emissions, Report Prepared for: California Energy Commission Public Interest Energy Research*, Mills, E., Lawrence Berkeley National Laboratory, 2009.
http://cx.lbl.gov/documents/2009-assessment/lbnl-cx-cost-benefit.pdf

8. SEE: Texas A&M Energy Systems Laboratory Study. 2009. http://esl.tamu.edu/

9. Data were provided by Servidyne Systems, LLC, the RCx service provider.

10. Data for RCx program results and case study were provided by Servidyne Systems, LLC.

11. Data were provided by Servidyne Systems, LLC.

12. *Performance Tracking: A Key Part of Ongoing Commissioning, Presentation for Energy Star Monthly Partner Web Conference*, Moser, D., Portland Energy Conservation, Inc., August, 2010.
http://www.energystar.gov/ia/business/networking/
presentations_2010/Aug10_On-Going_Commissioning.pdf

13. *Building Commissioning—A Golden Opportunity for Reducing Energy Costs and Green-house Gas Emissions, Report Prepared for: California Energy Commission Public Interest Energy Research*, Mills, E., Lawrence Berkeley National Laboratory, 2009.
http://cx.lbl.gov/documents/2009-assessment/lbnl-cx-cost-benefit.pdf

Chapter 6

Standards, Codes & Rating Systems

One of the most difficult parts of selecting a path for high-performance is navigating the variety of standards, codes and rating systems vying for everyone's attention. Further complicating the endeavor is the fact that nearly all of these are undergoing substantial change on a regular basis. How do we know what is absolutely essential now, what we can expect building codes to mandate in the future, or what would simply be prudent to do from the standpoint of cost savings and marketability? There is no question that aside from what is already required by code, there is competition in the "marketplace" for standards and ratings as there is no single organization which has the last word. Therefore some background on these three areas will be helpful to clearly define their roles. We proceed to characterize and discuss all three categories as they are concerned with high-performance and outline the mandatory and voluntary, the prescriptive and performance paths.

STANDARDS

In general, building standards are created and promulgated by a National Standards Body (NSB), a National Standards Organization (NSO) or International Standards Organization (ISO). These organizations research and develop standards for building practice and construction materials using technical advisory groups or technical committees typically drawn from members of industry and academia, so that peer-review is an important component of scientific study. These standards are usually offered for adoption by building professionals and building code developers as well as local and national governments and are therefore typically offered on a voluntary basis until they become part of the law or the building code. Standards can be prescriptive only or they can also have a performance component which specifies minimum requirements for compliance.

American National Standards Institute (ANSI)

ANSI Uses Technical Advisory Groups (TAGs) to create and develop standards for buildings and building components. The Institute oversees the creation, promulgation and use of thousands of norms and guidelines that directly impact businesses in nearly every sector: from acoustical devices to construction equipment, from dairy and livestock production to energy distribution, and many more. ANSI is also actively engaged in accrediting programs that assess conformance to standards—including globally recognized cross-sector programs such as the ISO 9000 (quality) and ISO 14000 environmental management systems. Although to date ANSI has not developed any standards specifically related to high-performance building, it collaborates with standards organizations in that area, specifically the High Performance Building Council of the National Institute of Standards and Technology (NIST).[1]

American Society of Testing Materials (ASTM)

ASTM's building standards are instrumental in specifying, evaluating, and testing the dimensional, mechanical, rheological, and other performance requirements of the materials used in the manufacture of main and auxiliary building parts and components. These materials include construction sealants, structural members, insulation systems, and other facilities used in conjunction with the erection of the foundation, walls, roofs, ceilings, doors, and windows of both commercial and domestic building structures. Technical committees develop and maintain ASTM standards. ASTM committees are made up of over 32,000 volunteers from industry and include manufacturers and consumers, as well as other interest groups such as government or academia. These building standards are helpful in guiding manufacturers, construction companies, architectural firms, and other users of such parts and components in their proper fabrication and installation procedures, as well as the possible hazards involved during these processes.[2] ASTM publishes dozens of standards which can be applied to high-performance buildings.

National Institute of Standards & Technology (NIST)

Founded in 1901, NIST is a non-regulatory federal agency within the U.S. Department of Commerce. NIST's mission is to promote U.S. innovation and industrial competitiveness by advancing measurement science, standards, and technology in ways that enhance economic security and improve our quality of life. NIST operates under a Memorandum of Understanding with ANSI with regard to the creation and promulgation of standards. From automated teller machines and atomic clocks to mammograms and semiconductors, innumerable products and services rely in some way on technology, measurement, and standards provided by the National Institute of Standards and Technology. NIST employs about 2,900 scientists, engineers, technicians, and support and administrative personnel. Also, NIST hosts about 2,600 associates and facility users from academia, industry, and other government agencies. In addition, NIST partners with 1,600 manufacturing specialists and staff at about 400 MEP service locations around the country.[3]

NIST is engaged in the development of **Measurement Science for Net-Zero Energy, High-Performance Buildings**[4], and specifies the problem as follows:

> *Buildings are complex systems of interacting subsystems. Past improvements in the energy performance of individual components/systems have not resulted in the expected reductions in overall building energy consumption. The industry is very sensitive to the first cost of new technologies, and performance goals, such as energy efficiency, indoor air quality, and comfort often conflict. Because a mismatch exists between who invests (builders and manufacturers) and who benefits (public), public sector involvement is necessary to overcome the initial barrier of developing the measurement science. Performance measurements made on individual components in carefully controlled laboratory test environments are idealized and capture neither the complexities of actual building installation nor the dynamic interactions of multiple subsystems. An integrated portfolio of measurement science capabilities is needed that not only supports innovation in the design and manufacturing of individual components and systems, but also captures the system complexities and interactions seen in a real building. Each individual measurement capability presents technical challenges, and the overall goal of significantly improved energy performance can only be achieved by applying an integrated portfolio of such measurement science capabilities.[5]*

NIST's goal is to use its extensive experience and testing facilities to help validate the building science and provide the measurement techniques and software applications necessary for achieving net zero, high performance buildings.

National Institute of Building Sciences (NIBS)

NIBS was authorized by the U.S. Congress in the Housing and Community Development Act of 1974, and is a non-profit, non-governmental organization bringing together representatives of government, the professions, industry, labor and consumer interests to focus on the identification and resolution of problems and potential problems that hamper the construction of safe, affordable structures for housing, commerce and industry throughout the United States. In establishing the Institute, Congress recognized the need for an organization that could serve as an interface between government and the private sector. The Institute's public interest mission is to serve the Nation by supporting advances in building science and technology to improve the built environment. Through the Institute, Congress established a public/private partnership to enable findings on technical, building-related matters to be used effectively to improve government, commerce and industry. The Institute provides an authoritative source of advice for both the private and public sector of the economy with respect to the use of building science and technology. Since 1997, the NIBS has been developing the Whole Building Design Guide (WBDG), a web-based, open access portal for government and industry practitioners in association with (among others) the US Air Force, the Army Corps of Engineers, The Department of Energy, NASA and the Sustainable Buildings Industry Council. Data from the U.S. Energy Information Administration illustrate that buildings are responsible for almost half (48%) of all greenhouse gas emissions annually and seventy-six percent of all electricity generated by U.S. power plants goes to supply the building sector.[6]

The WBDG states the following:

Figure 6-1

"Whole Building Design consists of two components: an in-

tegrated design approach and an *integrated team process*. The *"integrated"* design approach asks all the members of the building stakeholder community, and the technical planning, design, and construction team to look at the project objectives, and building materials, systems, and assemblies from many different perspectives. This approach is a deviation from the typical planning and design process of relying on the expertise of specialists who work in their respective specialties somewhat isolated from each other.

Whole Building design in practice also requires an integrated team process in which the design team and all affected stakeholders work together throughout the project phases and to evaluate the design for cost, quality-of-life, future flexibility, efficiency; overall environmental impact; productivity, creativity; and how the occupants will be enlivened. The 'Whole Buildings' process draws from the knowledge pool of all the stakeholders across the life cycle of the project, from defining the need for a building, through planning, design, construction, building occupancy, and operations. To create a successful high-performance building, an interactive approach to the design process is required. It means all the stakeholders—everyone involved in the planning, design, use, construction, operation, and maintenance of the facility—must fully understand the issues and concerns of all the other parties and interact closely throughout all phases of the project."[7]

This approach is quite obviously in consonance with integrated systems and concurrent engineering in manufacturing and is the most fully developed of the standards to date which encompasses "holistic" building practice.

NIBS formed the High Performance Building Council in 2007 in response to Section 914 of the 2005 Energy Policy Act. NIBS is working with the U.S. Department of Energy and U.S. standards writing organizations to assess current voluntary consensus standards and rating systems to provide for high performance buildings. In creating the High Performance Building Council, the overall goal is to put standards in place to define the performance goals of a high performance building in order to facilitate the design, construction, financing, and operation of buildings with an emphasis on life cycle issues rather than initial costs. The council will identify the metrics and level of required performance for specific design objectives (energy, security, durability, moisture, acoustics, etc) for building products, subsystems and systems; and reference industry standards for validating these performance requirements.[8]

International Organization for Standardization (ISO)

ISO is the oldest standards organization in the world. ISO was born from the union of two organizations—the ISA (International Federation of the National Standardizing Associations), established in New York in 1926, and the UNSCC (United Nations Standards Coordinating Committee), established in 1944. In October 1946, delegates from 25 countries, meeting at the Institute of Civil Engineers in London, decided to create a new international organization, of which the object would be "to facilitate the international coordination and unification of industrial standards." The new organization, ISO, officially began operations on 23 February 1947. ISO is the world's largest standards developing organization. Between 1947 and the present day, ISO has published more than 18,500 international standards, ranging from standards for activities such as agriculture and construction, through mechanical engineering, to medical devices, to the newest information technology developments.

ISO standards are developed according to the following principles.

* *Consensus*
  The views of all interests are taken into account: manufacturers, vendors and users, consumer groups, testing laboratories, governments, engineering professions and research organizations.

* *Industrywide*
  Global solutions to satisfy industries and customers worldwide.

* *Voluntary*
  International standardization is market driven and therefore based on voluntary involvement of all interests in the marketplace.

ISO publishes Standards 140001, Environmental Management, and 50001, Energy Management which are prescriptive, not performance based. These standards do not specify levels of environmental or energy performance. The intention of ISO 14001:2004 is to provide a framework for a holistic, strategic approach to an organization's environmental policy, plans and actions. ISO 14001:2004 gives the generic requirements for an environmental management system. The underlying philosophy is that whatever the organization's activity, the requirements of an effective EMS are the same. This has the effect of establishing a common reference for communicating about environmental management issues between organizations and their customers, regulators, the public and other stakeholders.[9]

Like all ISO management system standards, ISO 50001 has been designed for implementation by any organization, whatever its size or activities, whether in public or private sectors, regardless of its geographical location. ISO 50001 does not fix targets for improving energy performance. This is up to the user organization, or to regulatory authorities. This means that any organization, regardless of its current mastery of energy management, can implement ISO 50001 to establish a baseline and then improve on this at a rhythm appropriate to its context and capacities. The request to ISO to develop an international energy management standard came from the United Nations Industrial Development Organization (UNIDO) which had recognized industry's need to mount an effective response to climate change and to the proliferation of national energy management standards. Experts from the national standards bodies of 44 ISO member countries participated within ISO/PC 242 in the development of ISO 50001.[10]

As the US is a longstanding member of ISO, it is highly likely standards being developed for high-performance will be referenced or become part of a compliance or prescriptive path in ISO 50001 and 140001 in the near future.

American Society of Heating, Refrigerating & Air-Conditioning Engineers (ASHRAE)

The American Society of Heating, Refrigerating and Air Conditioning Engineers (ASHRAE,) is an international technical society founded in 1894 for all individuals and organizations interested in heating, ventilation, air-conditioning, and refrigeration (HVAC&R). The society, organized into regions, chapters, and student branches, allows exchange of HVAC&R knowledge and experiences for the benefit of the field's practitioners and the public. Technical committees meet typically twice per year at the ASHRAE annual and winter conferences.

ASHRAE's mission is "To advance the arts and sciences of heating, ventilating, air conditioning and refrigerating to serve humanity and promote a sustainable world. ASHRAE will be the global leader, the foremost source of technical and educational information, and the primary provider of opportunity for professional growth in the arts and sciences of heating, ventilating, air conditioning and refrigerating."[11]

ASHRAE publishes well recognized standards and guidelines relating to HVAC systems and issues. These standards are often referenced in building codes, and are considered useful standards for use by consulting engineers, mechanical contractors, architects, and government agencies.

Examples are:

- Standard 55—*Thermal Environmental Conditions for Human Occupancy*
- Standard 62.1—Ventilation for Acceptable *Indoor Air Quality*
- Standard 62.2—Ventilation and Acceptable Indoor Air Quality in Low-Rise Residential Buildings
- **Standard 90.1**—Energy Standard for Buildings Except Low-Rise Residential Buildings—The Illuminating Engineering Society of North America (IESNA) is a joint sponsor of this standard.
- **Standard 189.1**—Standard for the Design of High Performance, Green Buildings Except Low-Rise Residential Buildings

We shall be concerned here with Standard 90.1 and Standard 189.1 as they bear on the subject of high performance. ASHRAE has essentially taken the lead with these standards that are prescriptive and performance-based to develop the methodologies and measurement systems required to evaluate the components of high-performance buildings.

Standard 90.1

ASHRAE 90.1 has been in use for over 35 years and since 2001 has been updated every 3 years, with 90.1-2010 as the latest version. In 1992 the EPA began to require the DOE to review early versions of 90.1 and require all states to have energy codes substantially equal to the latest "approved" version. ASHRAE 90.1 is the basis for the International Energy Conservation Code (IECC) promulgated by the International Code Council (ICC), and the 2009 IECC includes 90.1-2007 as a compliance option. The standard provides minimum energy-efficiency requirements for buildings other than low-rise residential buildings and that apply to the following elements of buildings: (1) new buildings and their systems, (2) new portions of buildings and their systems, (3) new systems and equipment in existing buildings, (4) new equipment or building systems specifically identified in the standard that are part of industrial or manufacturing processes.[12]

ASHRAE 90.1 has sections including Administration & Enforcement, Building Envelope, Heating, Ventilating & Air Conditioning, Service Water Heating, Power, Lighting, Electrical, Energy Cost Budget Method, Normative References including Appendices covering Rated R-Value of Insulation and Assembly U-Factor, C-Factor, and F-Factor determinations, Building Envelope Climate Criteria, Climate Data, etc.

It is important to state that although the 90.1 Standard is comprehensive in scope and provides methods for specifying the performance characteristics of all building components, it is not a method for performing whole-building energy analysis. Nor does it deal with issues of sustainability or environmental impact. It does specify how computer simulations and building energy analysis should be done to establish design energy cost and energy cost budget.[13]

Standard 189.1

In 2008 ASHRAE decided to integrate sustainability into its charter. In association with the United States Green Building Council (USGBC), ICC, IES, and ANSI (2010), it released **Standard 189.1-2009**—Standard for the Design of High-Performance Green Buildings in December of 2009.

"Standard 189.1 provides a "total building sustainability package" for those who strive to design, build and operate green buildings. It is designed to increase energy performance over Standard 90.1-2007 by an average of 30%. From site location to energy use to recycling, this standard sets the foundation for green buildings by addressing site sustainability, water use efficiency, energy efficiency, indoor environmental quality, and the building's impact on the atmosphere, materials and resources. Standard 189.1 serves as a jurisdictional compliance option to the Public Version 2.0 of the International Green Construction Code™ (IgCC) published by the International Code Council. The IgCC regulates construction of new and remodeled commercial buildings."[14] The purpose of Standard 189.1 is to "provide minimum requirements for the siting, design, construction, and plan for operations of high-performance green buildings to: (a) balance environmental responsibility, resource efficiency, occupant comfort and well-being, and community sensitivity, and (b) support the goal of development that meets the needs of the present without compromising the ability of future generations to meet their own needs."[15] The last phrase (b), very significantly, is a direct quote of the definition of Sustainability from the Brundtland Report, this being the first instance of such reference in a standard from an organization ostensibly devoted entirely to engineering and the metrics of energy efficiency.

Standard 189.1 concerns itself with the following areas: administration and enforcement, site sustainability, water use efficiency, energy efficiency, indoor environmental quality, the building's impact on the atmosphere, materials and resources, construction and plans for operation, normative references. Appendices include: Prescriptive Building Envelope Tables, Prescriptive Continuous Air Barrier, Prescriptive Equipment Efficiency Tables, Performance Options for Energy Efficiency, Indoor Air Quality Limits for Office Furniture Systems and Seating, Building Concentrations, Informative References, and Integrated Design. It is not within our scope here to review all of these sections, or even any of them in any detail as 189.1 is a 128-page document. But to illustrate the breadth and penetration of the Standard, we select two examples. The first is from Water Management.

Plumbing Fixture Maximum Volume
Water Closets (Toilets)

•	Flushometer Valve Type Single Flush	1.28 gal (4.8 L)
•	Flushometer Valve Type Effective Dual Flush	1.28 gal (4.8 L)
•	Tank-Type Single Flush and WaterSense-Certified	1.28 gal (4.8 L)
	Effective Dual Flush and WaterSense-Certified	1.28 gal (4.8 L)
Urinals		0.5 gal (1.9 L)
Faucets		
•	Public Lavatory	0.5 gpm (1.9 L/min)
•	Public Metering Self-Closing	0.25 gal (1.0 L) per Metering Cycle
•	Residential Bathroom Lavatory Sink and WaterSense-Certified	1.5 gpm (5.7 L/min)
•	Residential Kitchen Showerheads	2.2 gpm (8.3 L/min)
•	Residential	2.0 gpm (7.6 L/min)

- Residential Shower Compartment
  (Stall) in Dwelling Units and Guestrooms
  All Shower Outlets   2.0 gpm (7.6 L/min)

Subsystem Submetering Threshold
Cooling Towers (Meter On Makeup Water and Blowdown)
Cooling Tower Flow Through Tower   >500 gpm (30 L/s)
Evaporative Coolers Makeup Water   >0.6 gpm (0.04 L/s)
Steam and Hot Water Boilers   >500,000 Btu/h (50 kW) Input
Total Irrigated Landscape Area
    with Controllers   >25,000 ft^2 (2500 m^2)
Separate Campus or Project Buildings
    Consumption   >1,000 gal/day (3800 L/day)
Separately Leased or Rental Space
    Consumption   >1,000 gal/day (3800 L/day)
Any Large Water-using Process
    Consumption   >1,000 gal/day (3800 L/day)

These are fairly aggressive requirements, and there is clearly a major emphasis on waste reduction in water usage, which in recent years has become as important as waste reduction in energy usage. The section on site sustainability has similarly aggressive goals with regard to low water usage.

Since ASHRAE 189.1 is intended to exceed 90.1-2007's energy performance standard by approximately 30%, there are a variety of mechanisms in place to accomplish that. First of all, adherence to 189.1-2009 means meeting the prescriptive requirements of 90.1-2007, sections 5.4, 6.4, 7.4, 8.4, 9.4 and 10.4. Although there are, of course exceptions, the building project's design must show allocated space for installation of on-site renewable energy systems. Under Section 7.3.3, Energy Consumption Management, there is provision for energy monitoring as such:

"**Consumption Management**. Measurement devices with remote communications capability shall be provided to collect energy consumption data for each energy supply source to the building, including gas, electricity and district energy...The measurement devices shall have the capability to automatically communicate the energy consumption data to a data acquisition system."

Energy accountability built into the system: an excellent program for tracking the data and evaluating how "high-performance" is actually performing.

It is clear that ASHRAE's development intends that 90.1 and 189.1 will form the pillars of energy, high-performance and green building standards for many years to come, and a graphical representation of this evolution (as they see it) is shown in Figure 6-2.[16]

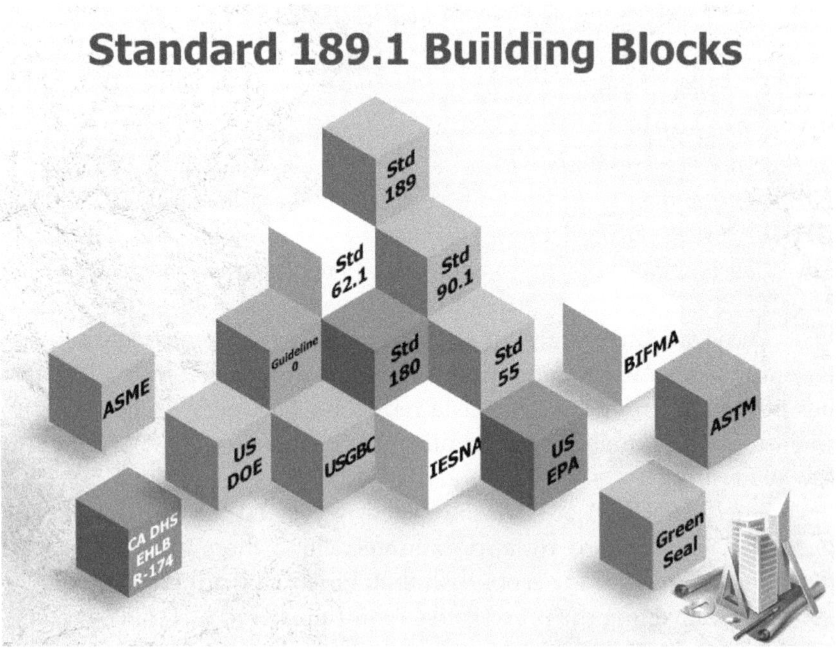

Standard 189.1 Building Blocks

Figure 6-2. Standard 189-1 Building Blocks

Standard 189.1 provides for both prescriptive and performance paths when evaluating compliance, so it allows for some flexibility with adoption, a key element in "selling" standards to the jurisdictional authorities such as states and municipalities. These last must obviously contend with market conditions bearing on the increased costs of mandating stricter code requirements for buildings and their performance. Nevertheless, the most important consideration is still that we have methods of validating the performance in operation with the performance that's been designed.

CODES

Building codes are designed and promulgated for adoption by reference in ordinance by authorities such as the federal government, states and municipalities, and are mandatory. "Since the early 1900s, the system of building regulations in the United States was based on model building codes developed by three regional model code groups. The codes developed by the Building Officials Code Administrators International (BOCA) were used on the East Coast and throughout the Midwest of the United States, while the codes from the Southern Building Code Congress International (SBCCI) were used in the Southeast and the codes published by the International Conference of Building Officials (ICBO) covered the West Coast and across to most of the Midwest. Although regional code development has been effective and responsive to the regulatory needs of the local jurisdictions, by early 1990s it became obvious that the country needed a single coordinated set of national model building codes. The nation's three model code groups decided to combine their efforts and in 1994 formed the International Code Council (ICC) to develop codes that would have no regional limitations. After three years of research and development, the first edition of the International Building Code was published in 1997."[17]

ICC's goal has been to develop a single set of comprehensive and coordinated national model construction codes. These model codes have now been adopted by 50 states and the District of Columbia at the state and jurisdictional level for mandatory compliance. The ICC publishes two codes related to energy efficiency and high-performance buildings which we shall consider below.

International Energy Conservation Code (IECC)

"The IECC is a model code that regulates minimum energy conservation requirements for new buildings and is revised every three years. The IECC addresses energy conservation requirements for all aspects of energy use in both commercial and residential construction including heating and ventilating, lighting, water heating, and power usage for appliances and building systems. The IECC is a design document."[18] It is used to design minimum insulation R-values and fenestration U-factors for the building envelope. The IECC sets forth minimum requirements for exterior envelope insulation, window and door U-factors and solar heat gain coefficient (SHGC) ratings[19], duct insulation, lighting and power efficiency, and water dis-

tribution insulation. ASHRAE 90.1-2007 is a compliance option for the 2009 IECC, the most recent version.

The IECC is divided up into six sections: Administration, Definitions, Climate Zones, Residential Energy Efficiency, Commercial Energy Efficiency, and Referenced Standards. An example of how the IECC is applied is as follows:

The code requirements vary by region. The regions are determined based on the climate and, hence, are called "climate zones." Table 301.1 lists Climate Zones, Moisture Regimes, and Warm-Humid Designations by state, county and territory. Figure 6-3 is a climate zone map outlining these regions and showing divisions by county as well.

The designer takes this information and can then design to meet the minimum requirements in the code by using tables and the prescriptive requirements set out in each section.

For example, for commercial buildings and residential buildings over three stories in height the building must comply with the energy efficiency requirements set forth in Chapter 5 of the 2009 IECC® or ASHRAE/IESNA 90.1.

Section 506 of the IECC provides criteria for compliance using total building performance, and includes the following systems and loads: heating systems, cooling systems, service water heating, fan systems, lighting power, receptacle loads and process loads. The idea here is that compliance can be achieved using a computer building simulation as long as the software tools are capable of calculating the annual energy consumption of all building elements, i.e. whole building energy analysis. As Section 506.6.1 reads: "Performance analysis tools meeting the applicable subsections of Section 506 and tested according to ASHRAE Standard 140[20] shall be permitted to be approved."

It is important to point out that not everything in the IECC-2009 is mandatory. There are some merely prescriptive elements allowing for flexibility in the application of the code.

International Green Construction Code (IgCC)

Outside of ASHRAE's Standard 189.1, the most important document to be published and which will ultimately be adopted by the building community in the area of high-performance buildings is the ICC's International Green Construction Code. The public version 2.0 was ratified by ICC in November, 2011 and the first edition was published in May, 2012. It will be revised and updated every three years.

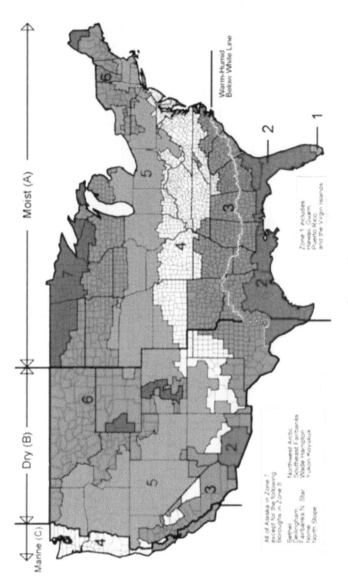

Figure 6-3

Table 6-1

Source: IECC 2009 TABLE 502.3
BUILDING ENVELOPE REQUIREMENTS: FENESTRATION

Climate Zone	1	2	3	4 Except Marine	5 and Marine 4	6	7	8
Vertical Fenestration (40% maximum of above-grade wall)								
U-Factor								
Framing materials other than metal with or without metal reinforcement of cladding								
U-Factor	1.2	0.75	0.65	0.4	0.35	0.35	0.35	0.35
Framing with or without thermal break								
Curtain Wall/Storefront U-Factor	1	0.7	0.6	0.5	0.45	0.45	0.4	0.4
Entrance Door U-Factor	1.2	1.1	0.9	0.85	0.8	0.8	0.8	0.8
All Other U-Facotr	1.2	0.75	0.65	0.55	0.55	0.55	0.45	0.45
SHGC-All Frame Types								
SHGC: PF < 0.25	0.25	0.25	0.25	0.4	0.4	0.4	0.45	0.45
SHGC: 0.25 < PF < 0.5	0.33	0.33	0.33	NR	NR	NR	NR	NR
SHGC > 0.5	0.4	0.4	0.4	NR	NR	NR	NR	NR
Skylights (3% Maximum)								
U-Factor	0.75	0.75	0.65	0.6	0.6	0.6	0.6	0.6
SHGC	0.35	0.35	0.35	0.4	0.4	0.4	NR	NR

a. All others includes operable windows, fixed windows and nonentrance doors.

It is not an overstatement to say that the IgCC will fundamentally transform the market for high-performance, sustainable buildings in the US. There are two principle reasons: (1) IgCC was created to be administered by code officials and adopted by governmental units on a *mandatory* basis, and (2) ASHRAE Standard 189.1 is included as an alternative compliance path. The IgCC was developed in association with ASTM, AIA and supported by USGBC and the Green Building Initiative (GBI)[21]. It applies to all buildings, new or existing, and alterations and additions, except low-rise residential buildings that fall under the scope of the International Residential Code (IRC)[22]. It will drive high-performance green building significantly beyond the market transformation which has already occurred with *voluntary rating systems* such as LEED, Green Globes[23], etc.

The IgCC is intended to be an overlay code which relies on the foundation of other international codes from ICC to produce buildings which are both safe and sustainable. "Rather than the past approach of creating buildings which are capable of resisting environmental forces, consideration is given in the IgCC to the impacts of building construction on the natural environment and how negative impacts can be mitigated. The IgCC, much like the IECC, is a code which regulates primarily from a *public welfare* perspective...The benefits of the IgCC are not only environmental. Because the IgCC approaches conservation from many perspectives, and conservation inherently means less materials, water and energy, etc..., in some scenarios, especially when considered over the useful life of the buildings and structures which conform to the IgCC, owners are likely to realize cost savings."[24]

The IgCC is not a rating system as it is composed primarily of mandatory requirements, shown in Table 6-2. But in addition to the jurisdictional requirements shown in Table 6-2, it also contains owner/designer choices called project electives, shown in Table 6-3.

A minimum number of these electives must be selected by the jurisdictional authority above and beyond the mandatory prescriptive and performance-based aspects of the code. The IgCC wants to encourage, but not mandate at this point, the voluntary implementation of practices which move toward *net-zero energy buildings*. This approach is contained in the project electives and allows designers, owners and code officials flexibility in reaching for that goal.

IgCC uses the concepts of zero Energy Performance Index (zEPI)[27], Energy Use Intensity (EUI)[28], and outcome-based compliance[29], to help

Table 6-2[25]

TABLE 302.1
REQUIREMENTS DETERMINED BY THE JURISDICTION

Section	Section Title or Description and Directives	Jurisdictional Requirements	
→			
CH 3. JURISDICTIONAL REQUIREMENTS AND PROJECT ELECTIVES			
302.1 (2)	Optional compliance path – ASHRAE 189.1	☐ Yes	☐ No
302.1 (3)	Project Electives – The jurisdiction shall indicate a number between 1 and 14 to establish the minimum total number of project electives that must be satisfied.	____	
CH 4. SITE DEVELOPMENT AND LAND USE			
→			
402.2.3	Conservation area	☐ Yes	☐ No
402.2.5	Agricultural land	☐ Yes	☐ No
402.2.6	Greenfields	☐ Yes	☐ No
402.3.2	Stormwater management	☐ Yes	☐ No
403.4.1	High occupancy vehicle parking	☐ Yes	☐ No
403.4.2	Low emission, hybrid and electric vehicle parking	☐ Yes	☐ No
405.1	Light pollution control	☐ Yes	☐ No
CH 5. MATERIAL RESOURCE CONSERVATION AND EFFICIENCY			
→			
502.1	Minimum percentage of waste material diverted from landfills.	☐ 50% ☐ 65% ☐ 75%	
CH 6. ENERGY CONSERVATION AND EARTH ATMOSPHERIC QUALITY			
→			
Table 602.1, 302.1, 302.1.1	zEPI of Jurisdictional Choice – The jurisdiction shall indicate a zEPI of 46 or less in Table 602.1 for each occupancy for which it intends to require enhanced energy performance.	See Table 602.1 and Section 302.1	
602.3.2.3	Total annual CO2e emissions limits and reporting	☐ Yes	☐ No
613.2	Post Certificate of Occupancy zEPI, energy demand, and CO2e emissions reporting	☐ Yes	☐ No
CH 7. WATER RESOURCE CONSERVATION AND EFFICIENCY			
→			
702.1.2	Enhanced plumbing fixture and fitting flow rate tier.	☐ Tier 1 ☐ Tier 2	
702.7	Municipal reclaimed water.	☐ Yes	☐ No
CH 9. COMMISSIONING, OPERATION AND MAINTENANCE			
904.1.1.1	Periodic reporting	☐ Yes	☐ No

Table 6-2[25]

Section	Section Title or Description and Directives	Jurisdictional Requirements	
	CH 10. EXISTING BUILDINGS		
→ 1006.4	Evaluation of existing buildings	☐ Yes	☐ No
	APPENDICES		
Appendix B	Greenhouse gas reduction in existing buildings	☐ Yes	☐ No
B103.1	Compliance level – The jurisdiction to select phases only where "Yes" is selected in the previous row.	☐ Phase 1 ☐ Phase 2 ☐ Phase 3 ☐ Phase 4	
B103.2	Where "Phase 1" is selected under Section B103.1 – jurisdiction to indicate the number of months to be used in association with Section B103.2.	_____ months	
B103.3	Where "Phase 2" is selected under Section B103.1 – jurisdiction to indicate the number of years and the percentage to be used in association with Section B103.3.	_____ years _____ %	
B103.4	Where "Phase 3" is selected under Section B103.1 – jurisdiction to indicate the number of years to be used in association with Section B103.4.	_____ years	
B103.5	Where" Phase 4" is selected above – jurisdiction to indicate the number of years and the percentage to be used in association with Section B103.5.	_____ years _____ %	
Appendix C	Sustainability measures	☐ Yes	☐ No
Appendix D	Enforcement procedures	☐ Yes	☐ No

realize buildings which come closer to meeting their intended energy goals. For example, a zEPI of 100 is intended to be indicative of the median energy use for similar commercial buildings constructed in the year 2000. Because studies have indicated that buildings constructed in accordance with the 2006 IECC consume approximately 27 percent less energy than buildings constructed and operated circa 2000, Section 602.2.1 deems buildings constructed in accordance with the 2006 IECC to have a zEPI of 73 (27 percent less than 100). As IGCC Public Version 2.0 is intended to require energy performance that is at least 30 percent better than buildings constructed in accordance with the *2006 IECC*, a zEPI of 51 (30 percent less than 73) is deemed the "point of entry" to the provisions of IGCC Chapter 6 as indicated in Table 602.1.[30]

Table 6-3[26]

TABLE 303.1
PROJECT ELECTIVES CHECKLIST

Section	Description	Check the corresponding box to indicate each project elective selected.	Jurisdictional determination of non-availability
CH 3. JURISDICTIONAL REQUIREMENTS AND PROJECT ELECTIVES			
304.1	Whole Building Life Cycle Assessment (LCA)	☐ (5 Electives[a])	☐
CH 4. SITE DEVELOPMENT AND LAND USE			
407.2.1	Flood hazard avoidance	☐	☐
407.2.2	Agricultural land	☐	☐
407.2.3	Wildlife corridor	☐	☐
407.2.4	Infill site	☐	☐
407.2.5	Brownfield site	☐	☐
407.2.6	Existing building reuse	☐	☐
407.2.7	Greenfield development	☐	☐
407.2.8	Greenfield proximity to development	☐	☐
407.2.9	Greenfield proximity to diverse uses	☐	☐
407.2.10	Native plant landscaping	☐	☐
407.2.11	Site restoration	☐	☐
407.3.1	Changing and shower facilities	☐	☐
407.3.2	Long term bicycle parking and storage	☐	☐
407.3.3	Preferred parking	☐	☐
407.4.1	Site hardscape 1	☐	☐
407.4.2	Site hardscape 2	☐	☐
407.4.3	Site hardscape 3	☐	☐
407.4.4	Roof covering	☐	☐
407.5	Light pollution	☐	☐
CH 5. MATERIAL RESOURCE CONSERVATION AND EFFICIENCY			
508.2	Waste management (502.1 + 20%)	☐	☐
508.3(1)	Reused, recycled content, recyclable, bio-based and indigenous materials (70%)	☐	☐
508.3(2)	Reused, recycled content, recyclable, bio-based and indigenous materials (85%)	☐ (2 Electives[a])	☐
→			
508.4.1	Service life – 100 year design life category	☐	☐
508.4.1	Service life– 200 year design life category	☐ (2 Electives[a])	☐

Table 6-3[26]

Section	Description	Check the corresponding box to indicate each project elective selected.	Jurisdictional determination of non-availability
508.4.2	Interior adaptability	☐	☐
→			
CH 6. ENERGY CONSERVATION, EFFICIENCY AND EARTH ATMOSPHERIC QUALITY		
613.3	zEPI reduction project electives		
613.3	Project zEPI is at least 5 points lower than required by Table 302.1	☐	☐
613.3	Project zEPI is at least 10 points lower than required by Table 302.1	☐ (2 electives)	☐
613.3	Project zEPI is at least 15 points lower than required by Table 302.1	☐ (3 electives)	☐
613.3	Project zEPI is at least 20 points lower than required by Table 302.1	☐ (2 electives)	☐
613.3	Project zEPI is at least 25 points lower than required by Table 302.1	☐ (4 electives)	☐
613.3	Project zEPI is at least 30 points lower than required by Table 302.1	☐ (5 electives)	☐
613.3	Project zEPI is at least 35 points lower than required by Table 302.1	☐ (6 electives)	☐
613.3	Project zEPI is at least 40 points lower than required by Table 302.1	☐ (8 electives)	☐
613.3	Project zEPI is at least 45 points lower than required by Table 302.1	☐ (9 electives)	☐
613.3	Project zEPI is at least 51 points lower than required by Table 302.1	☐ (10 electives)	☐
→			
613.4	Mechanical systems	☐	☐
613.5	Service Water Heating	☐	☐
613.6	Lighting Systems	☐	☐
613.7	Passive design	☐	☐
CH 7. WATER RESOURCE CONSERVATION AND EFFICIENCY		
710.2.1	Fixture flow rates are one tier above that required by Table 302.1	☐	☐
710.2.1	Fixture flow rates are two tiers above that required by Table 302.1.	☐ (2 Electives[a])	☐
710.3	On-site wastewater treatment	☐	☐
710.4	Non-potable outdoor water supply	☐	☐
710.5	Non-potable water for plumbing fixture flushing	☐	☐
710.6	Automatic fire sprinkler system	☐	☐
710.7	Non-potable water supply to fire pumps	☐	☐
710.8	Non-potable water for industrial process	☐	☐

Table 6-3[26]

Section	Description	Check the corresponding box to indicate each project elective selected.	Jurisdictional determination of non-availability
	makeup water		
710.9	Efficient hot water distribution system	☐	☐
710.10	Non-potable water for cooling tower makeup water	☐	
710.11	Graywater collection	☐	☐
	CH 8 INDOOR ENVIRONMENTAL QUALITY AND COMFORT		
809.2.1	VOC emissions – flooring	☐	☐
809.2.2	VOC emissions – ceiling systems	☐	☐
809.2.3	VOC emissions- wall systems	☐	☐
809.2.4	Total VOC limit	☐	☐
809.3	Views to building exterior	☐	☐
809.4	Interior plant density	☐	☐

a. Where multiple electives are shown in the table in the form "(x electives)", "x" indicates the number of credits to be applied for that elective to the total number of *project electives* required by the jurisdiction as shown in Section 302.1(3) of Table 302.1.

Section 604, Energy Metering, Monitoring and Reporting, requires all buildings that consume energy, regardless of compliance path, have capabilities for *energy measuring, monitoring* and reporting, or incorporate features that readily facilitate those capabilities in the future. The intent is to provide building owners, operation and maintenance staff with information which they can use to verify that buildings perform, and continue to perform in accordance with the IgCC.[31]

Another interesting and important project elective, 304.1, has to do with whole building life cycle assessment.[32] Life cycle analysis (LCA) and assessment are extraordinarily complex and many of the scientific and technical issues are still being worked out, but the IgCC Sustainable Building Technology Committee felt it should be included, but not be mandatory at this time. It is worth looking at the scope of the section to see what kinds of goals are involved:

Note that the last provision refers to the requirements of ISO 14044. It is clear that ultimately high-performance buildings will also need to include LCA as part of a holistic and sustainable approach to energy and environmental performance.

SECTION 304[33]
WHOLE BUILDING LIFE CYCLE ASSESSMENT

304.1 Whole building life cycle assessment project elective. A whole *building life cycle assessment* shall be a project elective. The requirements for the execution of a whole building life cycle assessment shall be performed in accordance with the following: The data and final report shall be included in the owner education manual required by Section 904.4.

1. The assessment shall demonstrate that the building project achieves not less than a 20 percent improvement in environmental performance for each of at least three of the following impact measures, one of which shall be *global warming potential*, as compared to a reference *building* of similar useable floor area, function and configuration that meets the minimum energy requirements of this code and the structural requirements of the *International Building Code*:

   1.1 *Primary energy use*
   1.2 Global warming potential
   1.3 Acidification potential
   1.4 Eutrophication potential
   1.5 Ozone depletion potential
   1.6 Smog potential

2. The reference and project *buildings* shall utilize the same life cycle assessment tool.

3. The *life cycle assessment* tool shall be approved by the code official.

4. *Building* operational energy shall be included.

5. *Building* process loads shall be permitted to be included.

6. The *reference service life* of the reference *building* shall be in accordance with Section 505.1.

7. Maintenance and replacement schedules and actions for components shall be included in the assessment.

8. The full life cycle, from resource extraction to demolition and disposal, including but not limited to, on-site construction, maintenance and replacement, and material and product embodied acquisition, process and transportation energy, shall be assessed.

   Exception: Electrical and mechanical equipment and controls, plumbing products, fire detection and alarm systems, elevators and conveying systems shall not be included in the assessment.

9. The complete *building* envelope, structural elements, inclusive of footings and foundations, and interior walls, floors and ceilings, including interior and exterior finishes, shall be assessed to the extent that data is available for the materials being analyzed in the selected *life* cycle assessment tool.

10. Mechanical, electrical and plumbing components and specialty items shall not be included in this calculation.

11. The *life cycle assessment* shall conform with the requirements of ISO 14044.

It is important to step back for a moment here in our review to comment on the parallel developments in the ASHRAE Standard 189.1 and the IgCC. *It is clear that for the first time with these two organizations, a direct connection is being made in the notion of high-performance building between significantly increasing energy efficiency and at the same time significantly reducing environmental impact through sustainable practices. This means that the environment external to the building itself has now become of major importance to the process.*

The IgCC will be the most comprehensive and detailed code available for the construction of high-performance sustainable buildings. Although the rate of adoption by jurisdiction is unknown, it will have a major impact on energy use and environmental impact by 2020.

RATING SYSTEMS

For the purposes of high-performance, a rating system is a program designed to improve the energy efficiency, environmental impact and sustainability of the built environment and/or the components and materials used in it, and is offered for adoption on a *voluntary* basis. Rating systems are sometimes based on standards and codes, but use these on a selected basis to achieve desired results which are typically beyond the scope of the existing regulatory environment. They have been used in the past as a way to drive market transformation without the force of the law. As such, they have been extremely valuable and will continue to be valuable as a way to set the bar as high as possible for the transformation of the built environment to high-performance. In the last ten to fifteen years, the regulatory environment has responded by gradually increasing the mandatory requirements in codes and standards to eventually achieve the very ambitious goals of net-zero energy, Cradle-to-Cradle[34] and the 2030 Challenge.[35]

Although rating systems are entirely voluntary, they still contain both prescriptive and performance-based requirements for compliance with the rating system. Typically the organization which sponsors the rating system will provide a "certification" resulting from a detailed process of assessment by a third party. The difference between standards and codes, and rating systems, can be in the application, as for example in the case of LEED, where a minimum threshold of requirements may be specified with a "point system" that allows the designer the option of achieving higher "ratings" above the minimum. This approach has sometimes been criticized for allowing too much flexibility and overemphasizing environmental impact at the expense of measurable energy performance and even financial performance. This criticism has some validity and the respective rating organizations have begun to address it. On the other hand, the metrics for predicting energy and financial performance are very complex and are still in development. Guaranteeing how buildings are going to perform is still a somewhat risky business. In any case, rating systems have been essential to the overall process and have consistently moved the dialogue forward in the area of high-performance.

Energy Star

The US Environmental Protection Agency established Energy Star in 1992 as a voluntary labeling system. It is now a joint program of the EPA and the DOE which provides ratings for appliances, heating and cooling systems, electronics, imaging equipment, lighting, desktop computers, and new homes. The energy performance improvements are expected to be in the 20-30% range compared with standard equipment or conventional homes. Commercial buildings can be rated as Energy Star and recently multi-family buildings were added as well. Once again, the baseline for comparison is essential to the value of the improvement.

Energy Star has developed an energy performance rating system for a variety of commercial, institutional and manufactured buildings. With a free, online tool called Portfolio Manager[36], a building owner or manager can enter energy information into the program and benchmark the energy efficiency of the building on a score of 1 to 100. The database of existing buildings available for comparison is extensive and includes the energy performance of more than 200,000 buildings and over 20 billion sq. ft., or approximately 25% of the total market. This makes Portfolio Manager (PM) an excellent method for profiling buildings *before the path to high-performance is determined*. Benchmarking with PM is thus a relatively easy approach to preliminary evaluations and also allows for market comparisons. PM can help track energy and water use and carbon footprint.

Ratable space types as of November 2011 include: bank/financial institutions, courthouses, data centers, dormitories, hospitals, hotels, houses of worship, k-12 schools, medical offices, office buildings, retail stores, supermarkets, warehouses, wastewater treatment plants, data centers. The energy performance rating system compares the subject building to its peer group of buildings identified through the Commercial Building Energy Consumption Survey (CBECS)[37]. The CBECS[38] is a national survey conducted quadrennially by the DOE.

A building can be rated as Energy Star under the following guidelines, and can be applied for on-line:

Energy Performance Specification for Designing and Operating Buildings to Achieve ENERGY STAR®

The following specification is to indicate that a design project or an existing building meets U.S. Environmental Protection Agency (EPA) energy performance criteria to achieve EPA's ENERGY STAR. The design may be eligible to receive the ENERGY STAR certification graphic, which denotes that the estimated energy use is intended to be in the top 25 percent as compared to the U.S. building stock. Once the building is built and operating for at least 1 year, it may qualify to receive the ENERGY STAR plaque.

The Architect of Record (AOR) can apply to for the Designed to Earn the ENERGY STAR graphic from EPA for a specified project. The AOR must demonstrate that the final estimate of the building's energy use corresponds to a rating of 75 or better using the EPA's energy performance rating from the online *Target Finder* tool.

Specifying the 2030 Challenge Goal. Target Finder may also be used to confirm that a design project meets energy reduction goals specified in the 2030 Challenge. Projects that meet 50 percent energy reduction (from an average building) or higher typically exceed the ENERGY STAR threshold of 75 on the EPA rating scale; therefore, EPA encourages setting the target for commercial new construction projects at a higher level than 75. The language can be adapted to specify meeting both the 2030 goal and ENERGY STAR.

The EPA energy performance rating is derived from fuel consumption data of existing commercial buildings, which includes the total energy use associated with the buildings. Design energy targets must include all fuel sources and total estimated energy use for the building design. An incomplete design energy use profile could result in a high, but inaccurate, rating. Gaps in energy analysis must be addressed in order for the rating to be a useful indicator of future performance.

The building owner can apply for the ENERGY STAR plaque by demonstrating that, after at least 1 year of operation, the building energy consumption from utility bills must: 1) rate 75 or higher on. EPA's energy performance rating from the online *Portfolio Manager* tool; and 2) meet specific indoor environmental quality standards.[39]

Although Energy Star is not specifically a high-performance program, it could be considered a gateway to this approach and is therefore valuable to owners and managers as a preliminary measure.

Leadership in Energy & Environmental Design (LEED)
LEED was introduced in 2000 under the auspices of the United States Green Building Council (USGBC)(1998), a private organization, and is at this juncture the pre-eminent rating system for green building in the US. It provides third-party verification for buildings or communities which are built using strategies to improve performance in energy savings, water efficiency, CO_2 emissions, indoor environmental quality, and stewardship of resources and their impacts. LEED is a design tool, not a performance measurement tool, and has been well embraced by the architectural community. LEED version 3.0 was published in 2009. Since the year 2000, there have been more than 7,000 projects in the United States and 30 countries covering over 1.501 billion square feet (140 km²) of development area.

"LEED promotes sustainable building and development practices through a suite of rating systems that recognize projects that implement strategies for better environmental and health performance. The LEED rating systems are developed through an open, consensus-based process led by LEED committees, diverse groups of volunteers representing a cross-section of the building and construction industry. Key elements of the process include a balanced and transparent committee structure, technical advisory groups that ensure scientific consistency and rigor, opportunities for stakeholder comment and review, member ballot of new rating systems, and fair and open appeals."[40]

LEED certification is obtained after submitting an application documenting compliance with the requirements of the rating system as well as paying registration and certification fees. Certification is granted solely by the Green Building Certification Institute (GBCI), an independent non-profit that was established in 2008 with the support of USGBC, which is responsible for the third party verification of project compliance with LEED requirements. GBCI includes a network of ISO-compliant international certifying bodies, ensuring the consistency, capacity and integrity of the LEED certification process. The application review and certification process is handled in LEED Online, USGBC's web-based tool that employs a series of active PDF forms to automate filing documentation and communication between project teams and

GBCI's reviewers.

Buildings can qualify for four levels of certification:

- **Certified**—40 - 49 points
- **Silver**—50 - 59 points
- **Gold**—60 - 79 points
- **Platinum**—80 points and above
- The LEED for Homes rating system is different from LEED v3, with different point categories and thresholds that reward efficient residential design.

LEED promotes a whole-building approach to sustainability by recognizing performance in key areas:[41]

Sustainable Sites

Site selection and development are important components of a building's sustainability. The Sustainable Sites category discourages development on previously undeveloped land; seeks to minimize a building's impact on ecosystems and waterways; encourages regionally appropriate landscaping; rewards smart transportation choices; controls stormwater runoff; and promotes reduction of erosion, light pollution, heat island effect and construction-related pollution.

Water Efficiency

Buildings are major users of our potable water supply. The goal of the water efficiency category is to encourage smarter use of water, inside and out. Water reduction is typically achieved through more efficient appliances, fixtures and fittings inside and water-conscious landscaping outside.

Energy & Atmosphere

According to the U.S. Department of Energy, buildings use 39% of the energy and 74% of the electricity produced each year in the United States. The energy and atmosphere category encourages a wide variety of energy-wise strategies: commissioning; energy use monitoring; efficient design and construction; efficient appliances, systems and lighting; the use of renewable and clean sources of energy, generated on-site or off-site; and other innovative measures.

Materials & Resources

During both the construction and operations phases, buildings generate a lot of waste and use large quantities of materials and resources. The materials and resources category encourages the selection of sustainably grown, harvested, produced and transported products and materials. It promotes waste reduction as well as reuse and recycling, and it particularly rewards the reduction of waste at a product's source.

Indoor Environmental Quality

The U.S. Environmental Protection Agency estimates that Americans spend about 90% of their day indoors, where the air quality can be significantly worse than outside. The indoor environmental quality category promotes strategies that improve indoor air as well as those that provide access to natural daylight and views and improve acoustics.

Locations & Linkages

The LEED for Homes rating system recognizes that much of a home's impact on the environment comes from where it is located and how it fits into its community. The locations and linkages category encourages building on previously developed or infill sites and away from environmentally sensitive areas. Credits reward homes that are built near already-existing infrastructure, community resources and transit—in locations that promote access to open space for walking, physical activity and time outdoors.

Awareness & Education

The LEED for Homes rating system acknowledges that a home is only truly green if the people who live in it use its green features to maximum effect. The awareness and education category encourages home builders and real estate professionals to provide homeowners, tenants and building managers with the education and tools they need to understand what makes their home green and how to make the most of those features.

Innovation in Design

The innovation in design category provides bonus points for projects that use innovative technologies and strategies to improve a building's performance well beyond what is required by other LEED credits, or to account for green building considerations that are not specifically addressed elsewhere in LEED. This category also rewards projects for including a LEED Accredited Professional on the team to ensure a holistic, integrated approach to the design and construction process.

Regional Priority

USGBC's regional councils, chapters and affiliates have identified the most important local environmental concerns, and six LEED credits addressing these local priorities have been selected for each region of the country. A project that earns a regional priority credit will earn one bonus point in addition to any points awarded for that credit. Up to four extra points can be earned in this way. *See the Regional Priority Credits for your state.*

One important issue, however, regarding LEED buildings, has been the question of measuring performance on an on-going basis after construction. Data still need to be collected to verify the goals set out by the rating system and certified by the GBCI. "LEED Commissioning" might be a good way to approach it. At the same time, as we have already seen, the collection of data to verify the predicted energy performance of high-performance buildings is still very much under development and will require the concerted efforts of literally thousands of building owners and organizations over the next decade. Nevertheless, to address the after-the-fact energy performance questions, LEED is developing Advanced Energy Modeling for LEED, which uses ASHRAE 90.1 and DOE-2 based simulation programs (such as eQuest and Energy Plus)[42]. This will provide building energy performance reports (BEPS), building utility performance reports (BEPU), and energy cost summary reports (ES-D).

It is not an exaggeration to say that LEED, despite its shortcomings (extra costs, lack of energy performance metrics), has been largely instrumental in moving the discussion forward in the US regarding the importance of sustainable, high-performance buildings. Some of its most important concepts and directives have been incorporated into ASHRAE 189.1 and the IgCC, a testament to its essential and on-going influence.

Green Building Initiative (GBI)

The Green Building Initiative was instrumental in bringing the Green Globes environmental assessment and rating tool, which was developed in Canada, into the US in 2004. Green Globes essentially competes with LEED but has some features which distinguish it. In addition, GBI has been working on some important initiatives related to high-performance which merit discussion.

In association with ASHRAE, The GBI is working on the development of the Database for the Analysis of Sustainable High-Performance Buildings (DASH), a critical piece in the knowledge base required to evaluate predicted and actual energy performance. GBI has also teamed up with ANSI to produce a standard, officially named *ANSI/GBI 01-2010: Green Building Assessment Protocol for Commercial Buildings*. This standard was derived from the Green Globes environmental design and assessment rating system for new construction and was formally approved on March 24, 2010. The standard was developed following ANSI's highly regarded consensus-based guidelines. A variety of stakeholders including sustainability experts, architects, engineers, ENGOs, and industry groups participated in its development.

GBI touts the Green Globes rating system as less expensive and considerably easier to implement than LEED. It has an on-line assessment and performance tool which streamlines the process but it is also a point system like LEED. It has similar areas of focus:

Energy	Resources
Indoor Environment	Emissions
Site	Project/Environmental Management
Water	

Figure 6-4 is GBI's comparison of the two systems.

Is it possible to say whether Green Globes is a better system than LEED? Probably not, although it does have some advantages. Its flexibility with a choice of prescriptive or performance paths is a plus with the attendant energy performance assessment. Green Globes also has an LCA Credit Calculator which is based on the ATHENA LCA computer program. LEED is much more widely recognized and has considerably more market clout, but that does not mean that it is better either. At this point, they are two competing rating systems whose goals are the same: high-performance, sustainable buildings.

TRENDS & CONCLUSIONS

In the last several years, numerous municipalities (New York, Boston, San Francisco, Chicago, Washington, D.C., etc.) have either adopted or are in the process of adopting ordinances which *require all commercial*

Figure 6-4. Green Globes/LEED—A Comparison

Criteria	Green Globes	LEED
(ANSI) Developed Through Recognized Consensus Process	YES	NO
Nationally Accepted Environmental Rating and Assessment Program	YES	YES
Program Delivery	Web enabled interactive questionnaire	Online forms
New Construction Assessment	YES	YES
Existing Buildings Assessment	YES	YES
Program Points	1,000	110
Energy Performance (New Construction)	Benchmarks against actual regional performance data Benchmarks against hypothetical building model	
Criteria Weighted—Partial Credit Scores Possible	YES	LIMITED
Forest Certifications Accepted	4	1
Specific Prerequisite Items	NO	YES
Minimum Points Required for Certification	YES	YES
Incorporates Life Cycle Assessment	YES	NO
Flexibility for Non-Applicable Criteria	YES	NO
Automated Online Report Incorporates Sustainability Recommendations	YES	NO
Certification Process	Assessor assigned/on-site building audit with team	Fill out assessment form, submit, await results
Certification Ratings	4 Globes 3 Globes 2 Globes 1 Globe	LEED Platinum LEED Gold LEED Silver Certified LEED
Certified Personnel Training Program Available	YES	YES
Time Requirements to Complete Documentation	◓	●
Cost to Certify a Typical Building > 100,000 sq ft	$	$$$

Green Globes is North America's first web-enabled, fully interactive green building assessment tool that allows building professionals and owners to augment their design, in the case of new construction, or incorporate sustainability operations, in the case of existing buildings, and rate the building's proposed or actual sustainability performance. The system features allow building owners and managers to have first-hand knowledge at any given time how their building is scoring. If a building achieves at least 35% of the total number of 1,000 points, it qualifies for certification. Upon ordering the certification, a third-party assessor appointed by the GBI begins to work with the owner and team during the assessment period which culminates in an on-site audit of the building. Green Globes places an emphasis on benchmarking and improvements, providing an easier, affordable way to go green. Green Globes rating and certification process can be completed for a fraction of the combined hard/soft costs and time associated with LEED. Green Globes is ideal for complex or specialty buildings that cannot be certified with LEED.[43]

building owners(with some exemptions) to monitor and report their energy and water usage and make the information public. The emphasis so far has been on employing Energy Star's Portfolio Manager to benchmark existing buildings and letting the market drive competition for performance. Consider the following language from an ordinance passed in 2012 by the City of Minneapolis:

CHAPTER 47. ENERGY AND AIR POLLUTION

Section 2. That Chapter 47 of the Minneapolis Code of Ordinances be amended by adding thereto a new Section 47.190 to read as follows:

47.190. Commercial building rating and disclosure. *(a) Definitions. The following words shall have the meaning ascribed to them, unless the context clearly indicates a different meaning:*

Benchmark means to input the total energy consumed for a building and other descriptive information for such building as required by the benchmarking tool.

Benchmarking information means information related to a building's energy consumption as generated by the benchmarking tool, and descriptive information about the physical building and its operational characteristics. The information shall include, but need not be limited to:

(1) Building address;

(2) Energy use intensity (EUI);

(3) Annual greenhouse gas emissions;

(4) Water use; and

(5) The energy performance score that compares the energy use of the building to that of similar buildings, where available.

Benchmarking tool means the United States Environmental Protection Agency's Energy Star Portfolio Manager tool, or an equivalent tool adopted by the director.

Building owner means an individual or entity possessing title to a building, or an agent authorized to act on behalf of the building owner.

City-owned building means any building, or group of buildings on the same tax lot, owned by the City of Minneapolis containing 25,000 or more gross square feet of an occupancy use other than residential or industrial.

Covered building means:

(1) Any building containing at least 50,000 but less than 100,000 gross square feet of an occupancy use other than residential or industrial shall be classified as a Class 1 covered building;

(2) Any building containing 100,000 or more gross square feet of an occupancy use other than residential or industrial shall be classified as a Class 2 covered building.

The term "covered building" shall not include any building owned by the local, county, state, or federal government or other recognized political subdivision.

Director means the head of the department to which the environmental services division of the city reports or the director's designee.

Energy means electricity, natural gas, steam, heating oil, or other product sold by a utility for use in a building, or renewable on-site electricity generation, for purposes of providing heating, cooling, lighting, water heating, or for powering or fueling other end-uses in the building and related facilities.

Energy performance score means the numeric rating generated by the Energy Star Portfolio Manager tool or equivalent tool adopted by the director that compares the energy usage of the building to that of similar buildings.

Energy Star Portfolio Manager means the tool developed and maintained by the United States Environmental Protection Agency to track and assess the relative energy performance of buildings nationwide.

Tenant means a person or entity occupying or holding possession of a building or premises pursuant to a rental agreement.3

Utility means an entity that distributes and sells natural gas, electric, or thermal energy services for buildings.

(b) Benchmarking required for city-owned buildings. No later than May first, 2013, and no later than every May first thereafter, each city-owned building shall be benchmarked for the previous calendar year by the entity primarily responsible for the management of such building, in coordination with the director.

(c) Benchmarking required for covered buildings. Building owners shall annually benchmark for the previous calendar year each covered building and obtain an energy performance score as available according to the following schedule:

(1) All Class 2 covered buildings by May first, 2014 and by every May first thereafter; and

(2) All Class 1 covered buildings by May first, 2015 and by every May first thereafter.

(d) Disclosure and publication of benchmarking information. The building owner shall annually provide benchmarking information to the director, in such form as established by the director's rule, by the date provided by the schedule in subsections (b) and (c).

(1) The director shall make readily available to the public, and update at least annually, benchmarking information for the previous calendar year according to the following schedule:

a. Each city-owned building by July thirtieth, 2013 and by every July thirtieth thereafter;

b. Each Class 2 covered building by July thirtieth, 2015 and by every July thirtieth thereafter;

c. Each Class 1 covered building by July thirtieth, 2016 and by every July thirtieth thereafter.

(2) The director shall make available to the public, and update at least annually, the following information:

a. Summary statistics on energy consumption in city-owned buildings and covered buildings derived from aggregation of benchmarking information for those buildings;

b. Summary statistics on overall compliance with this section[44]

The direction is clear. Building owners and managers would be wise to get ahead of the curve in the regulatory environment rather than trying to play catch-up in the marketplace on down the road.

Also, developments in the last 15 years in standards, codes and rating systems show that there is confluence going on between the mandatory and voluntary approaches to achieving high-performance. Whereas there used to be building codes that were adopted and applied regionally and locally[45], it is clear that unification is occurring in model, national codes under the auspices of the ICC. This is a good development from the standpoint of high-performance, sustainable buildings, since then the scientific principles can be applied consistently and with measurement systems that have broad acceptance. The code officials, standards organizations and purveyors of rating systems are collaborating to produce a more unified approach. ASHRAE 189.1 in conjunction with IgCC will put us on a steady path to improvement and consistent feedback on those improvements. Standards and rating systems, which are typically for sale and thus compete in the market for acceptance, can drive the process forward faster than the regulatory environment which takes longer to implement. They can respond to changes in the energy and resource environment and respond accordingly. That is also a good thing, since net-zero, zero waste and carbon-neutral must be our ultimate goals in high-performance.

References

1. http://www.ansi.org/about_ansi/overview/overview.aspx?menuid=1
2. http://www.astm.org/Standards/building-standards.html
3. www.nist.gov
4. Although there are different types of Net-Zero Energy Buildings, "In general, a net-zero energy building produces as much energy as it uses over the course of a year." See: http://www1.eere.energy.gov/buildings/commercial/index.html
5. http://www.nist.gov/el/highperformance_buildings/upload/01NetZeroEnergy_web.pdf
6. http://www.wbdg.org/wbdg_approach.php
7. http://www.wbdg.org/wbdg_approach.php
8. http://www.nibs.org/index.php/hpbc/
9. http://www.iso.org/iso/iso_catalogue/management_and_leadership_stan-

dards/environmental_management/iso_14000_essentials.htm

10. http://www.iso.org/iso/iso_50001_energy.pdf

11. http://www.ashrae.org/aboutus

12. ANSI/ASHRAE/IESNA Standard 90.1-2010, Atlanta, GA, p. 4.

13. ANSI/ASHRAE/IESNA Standard 90.1-2010, Atlanta, GA, p. 90.

14. http://www.ashrae.org/publications/page/927

15. ANSI/ASHRAE/USGBC/IES Standard 189.1-2009, Atlanta, GA, p.4

16. http://www.ashrae.org/publications/page/927

17. http://bulk.resource.org/codes.gov/

18. 2009 International Energy Conservation Code, ICC, Country Club Hills, IL, p. v.

19. SHGC: "The SHGC is the fraction of incident solar radiation admitted through a window, both directly transmitted and absorbed and subsequently released inward. SHGC is expressed as a number between 0 and 1." http://www.efficientwindows.org/shgc.cfm

20. ANSI/ASHRAE 140-2007 /Standard 140-2007 -- Standard Method of Test for the Evaluation of Building Energy Analysis Computer Programs (ANSI Approved) : This standard specifies test procedures for evaluating the technical capabilities and ranges of applicability of computer programs that calculate the thermal performance of buildings and their HVAC systems. http://shop.iccsafe.org/ansi-ashrae-140-2007-standard-method-of-test-for-the-evaluation-of-building-energy-analysis-computer-programs-download.html

21. The Green Building Initiative will be described later under Rating Systems.

22. The International Residential Code is the general building code for residential construction promulgated by ICC.

23. Green Globes is the Rating System of the Green Building Initiative.

24. IgCC Public Version 2.0 Synopsis, p. 4., available at http://www.iccsafe.org/cs/IGCC/Documents/PublicVersion/IGCC_PV2_Synopsis.pdf

25. IgCC Public Version 2.0, p. 29, available at : http://www.vpmia.org/pdf/Codes/IGCC-PV2_PDF.pdf

26. IgCC Public Version 2.0, p. 52.

27. Though there are other factors that affect the value of zEPI, it is essentially calculated in accordance with Equation 6-3 as follows:

　　zEPI = 57 x (PD - RE - WE) / RD (Equation 6-3)

　　Where:

　　PD = Total annual energy delivered to the proposed design and consumed on site, as determined in accordance with Section 603

　　RE = Total annual energy savings from renewable energy derived on site

　　RD = Total annual energy used by a standard reference design, determined in accordance with Section 603

　　WE= Total annual energy savings from waste energy recovery

　　PD, RE, RD and WE must be expressed in consistent units of energy in accordance with Section 603.1.1.

　　See: IgCC Public Version 2.0 Synopsis, p. 20

28. EUI = TAE/SF (where TAE = Total annual energy projected to be consumed on site, including renewable energy; and SF = building gross square footage) (Section 602.2.4.1), IgCC Public Version Synopsis, p. 20.

29. This refers to Annual Net Energy Performance (ANEP), see IgCC Public Version 2.0, Section 603, p. 77.

30. IgCC Public Version 2.0, Synopsis, p. 21.

31. IgCC Public Version 2.0, Synopsis, p. 22.

32. Life Cycle Assessment: "To examine how much a product impacts the environ-

ment, it is necessary to account for all the inputs and outputs throughout the life cycle of that product, from its birth, including design, raw material extraction, material production, part production, and assembly, through its use, and final disposal." See: www.enviroliteracy.org.

33. IgCC Public Version 2.0, p. 32

34. Cradle to Cradle is a part of LCA: "It is a holistic economic, industrial and social framework that seeks to create systems that are not just efficient but essentially waste free." Lovins, L. Hunter (2008). Rethinking Production, in State of the World 2008, pp. 38–40.

35. The 2030 Challenge sets the bar very high, as its goal is to make the built environment Carbon Neutral by 2030, i.e. eliminating fossil-fuel energy and Greenhouse Gas Emissions (GHG) from the built environment. See: http://www.architecture2030.org/2030_challenge/the_2030_challenge

36. See: https://www.energystar.gov/istar/pmpam/

37. As of the Spring 2011, the CBECS was excised through budget cuts, but was reinstated in April 2013. The NIBS began a High-Performance Building Data Collection Initiative in May, 2011 to help pick up the slack. Additionally, the GBI, now in consort with the American Society of Heating, Refrigerating and Air-Conditioning Engineers (ASHRAE)—has been working for several years to develop their Database for Analyzing Sustainable and High Performance Buildings (DASH) (a beta version to be published in late 2011), a publicly available collection of data about building operation, maintenance and performance, especially for green, sustainable and high-performance buildings. Guided primarily by practitioners, DASH could now play a role in maintaining contact with industry professionals and identifying gaps in currently collected building data, thereby serving as an important tool for the NIBS initiative. www.newbuildings.org/. See also: www.gbapgh.org/content.aspx?ContentID=92.

38. http://www.eia.gov/consumption/commercial/

39. www.energystar.gov/ia/business/tools_resources/new_bldg_design/cbd_guidebook/DEES_Guide_Energy_Performance_Spec.pdf

40. http://www.usgbc.org/DisplayPage.aspx?CMSPageID=1988

41. http://www.usgbc.org/DisplayPage.aspx?CMSPageID=1989

42. These Building Energy Analysis (BEA) Programs will be reviewed in the next chapter.

43. http://www.thegbi.org/green-globes/green-globes-leed-green-building-certification.shtml

44. http://www.minneapolismn.gov/www/groups/public/@regservices/documents/webcontent/wcms1p-101277.pdf

45. Apparently the City of Chicago is still a holdout, adhering to its own Municipal Building Code.

Chapter 7

Envelope

Tony Robinson, MS
Dr. Paul Tinari, P.E.: SOUND

INTRODUCTION

The building envelope consists of the walls, roof and floor of the structure and their component systems. From the standpoint of energy, its principal functions are to:

[1] manage heat flows between the external environment and the interior space,
[2] admit and control daylight,
[3] mitigate sound transmission, and
[4] divert/control flows of air(wind) and water (rain & humidity).

It can be stated simply that *the envelope controls the energy equation* and is therefore one of the most important parts of the design or retrofit of a high-performance building. What do we mean by "controls the energy equation?" We mean that *the envelope is the mediator of all energy transfers between the external environment and the interior space*. The interior of the building gains and loses energy based on:

[1] the regional & local climatic conditions,
[2] the mediation of the envelope, and
[3] the activities of the occupants.

It is obvious that we have no control over (1) and limited control over (3), depending on the building use.

Mechanical heating, cooling, ventilating and electric lighting are, of course, systems designed to moderate internal conditions for the occupants which cannot be controlled adequately due to external climate, the inefficiency of the envelope and the diurnal and nocturnal cycles.

Until recently, from the standpoint of energy transfers, we have been trained to regard the envelope as a *barrier*. We are trying to keep unwanted energy out, remove unwanted energy once it is inside the building, or keep wanted energy inside the building which we absorb from the exterior or produce through mechanical and electrical processes (i.e. heating, cooling, artificial lighting). Traditionally, the approach was to *reject and remove*. But in high-performance buildings, although it still acts as a barrier, we must consider the envelope more as a *membrane* and a *collector*, which filters and transforms the energy it absorbs and collects into the appropriate use. This approach is to *absorb and translate*. From this standpoint, any energy which is absorbed or reflected by the envelope and is not converted to a useful form, is wasted energy. In the future, we may describe the performance of the envelope in terms of "conversion efficiency," just as we do now for example with electric motors or engines.

In high-performance building envelopes, we have to be primarily concerned with optimizing the energy equation by leveraging (1), the passive features of the envelope (principally geometry and orientation) and (2), the physical properties of its material components (e.g. absorptance, conductivity, resistance, emissivity, and transmittance)[1]. A well designed or retrofitted envelope substantially reduces the requirements for heating, ventilating and air-conditioning equipment and also reduces electrical lighting requirements with daylighting. As a corollary, this also reduces the requirements for on-site or distributed energy. The high-performance building envelope as *collector*, therefore leverages all of the energy assets available on the site to decrease or potentially eliminate the necessity for energy supplied remotely. Once again, the ultimate goal of the high-performance building must be *zero-net energy*, or a *net-energy producer* (producing more energy than it consumes).

The design and energy analysis of the building envelope can be complex and there is still much to be learned about it. Building materials and techniques are evolving to respond to developments in materials science and manufactured component systems. However, the principles of the energy transfers in the envelope are well understood and can be explained here to outline the goals of high performance. We will discuss those principles as well as some case studies of envelopes which embody high-performance characteristics. It is not our intention to provide a summary of envelope designs or describe the physics of energy transfers. Those topics have been discussed in great detail in other volumes.[2] What we want to do is point out how certain fundamental principles can be employed to maximize the energy flows in and out of the building.

EXTERNAL CONDITIONS

Climate

As shown below, the upper atmosphere of the earth is a type of envelope which mediates energy transfers due to solar radiation in and out of the biosphere[3] and helps regulate the climate. This produces the "greenhouse effect," and the principles of it have application with the building envelope.

This "greenhouse" effect is shown in Figure 7-1[4], with the energy inputs and transfers expressed in watts per square meter.

Solar radiation affects the building envelope in numerous ways, including energy transfers from ambient climate conditions to direct radiation on the surfaces of the envelope. See Figure 7-2.

Sun-path diagrams[5] (Figure 7-3) are often used to determine the overall exposure of the building envelope to solar radiation and can be developed for any latitude.

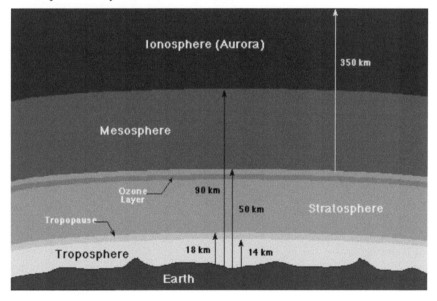

Figure 7-1

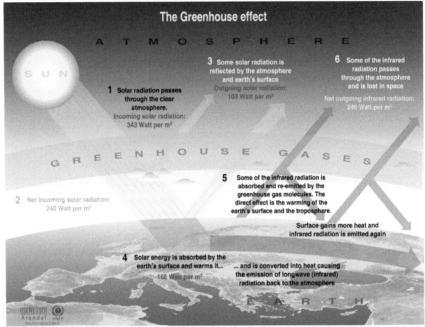

Sources: Okanagan university college in Canada, Department of geography, University of Oxford, school of geography; United States Environmental Protection Agency (EPA), Washington; Climate change 1995, The science of climate change, contribution of working group 1 to the second assessment report of the intergovernmental panel on climate change, UNEP and WMO, Cambridge university press, 1996.

Figure 7-2

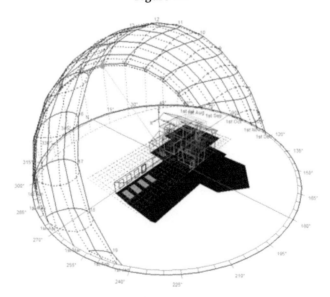

Figure 7-3

The envelope of a building can be viewed as an integrated system of materials which absorb, reflect and transmit the energy from solar radiation depending on their composition. Thus solid walls or roofs are opaque to the visible transmittance of solar energy but they reflect, absorb and transmit heat. Glazing, fenestration, windows, absorb, reflect and transmit solar radiation in the form of heat and visible light. A greenhouse (see Figure 7-3) is usually defined as a building with an envelope that is (apart from minimal structure) entirely glazed and can therefore be primarily transparent to visible light.[6]

A conventional building (i.e. that designed for continuous occupancy as in Figure 7-4)[7] has an envelope which combines solid walls, roof and glazing and so operates under some of the same principles of the greenhouse and the "greenhouse effect" in the earth's atmosphere. In other words, the building envelope, regardless of the percentage of glazing it contains, creates its own atmosphere and climate inside the building.

The performance of any building envelope is first of all determined by the regional climatic conditions[8], and also what we call the *external microclimate*. The external microclimate would include any of the following within a ¼-½ mile radius of the exterior of the building: bodies of water, air flows, rainfall, landforms (such as hills, mountains),

Figure 7-4

other buildings, trees, etc. In other words, things that would locally affect temperature, moisture, solar radiation on the building, heat islands, wind patterns, etc. External microclimatic conditions such as wind patterns can be modeled, for example, using wind roses, as shown in Figure 7-5.[9] High-performance strategies require careful consideration of local external microclimatic conditions in order to optimize energy transfers through the envelope. Since high-performance includes the idea of mitigating environmental impact, then it should be stated that we are not just concerned with how the external microclimate affects the building envelope, but *also how the building envelope affects the external microclimate*. (For example, a black roof surface radiating heat into the local atmosphere, or reflective glass from a curtain wall causing excessive glare or reflection on a nearby building).

Thus a building becomes an integrated physical system whose energy inputs and outputs can also significantly affect the other buildings in the area. In other words, unless you are designing a house in the middle of a 1000-acre ranch in Utah, then the envelopes of buildings in close proximity to one another, if they are to be high-performance, should not be treated as if they are independent physical systems (including residential buildings).

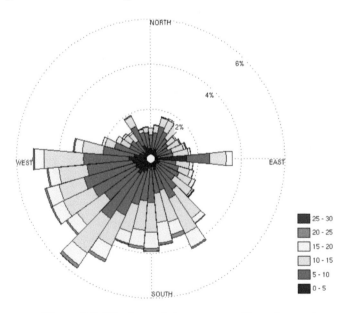

Figure 7-5. Wind rose (please see: mathworks.com)

Site, Orientation & Geometry

The siting and the orientation of the building (north-south, east-west, axes) are also critical to the performance of the envelope. If the main facade of your building faces west and you are in the southwestern US in the desert, then a 20-story glass curtain wall is going to gain a lot of heat energy in July. If the main facade of your building faces east and you are in the northeastern US on the ocean, then a 20-story glass curtain wall is going to lose a lot of heat energy in December. So energy loads are not only climate specific, they are also site specific and orientation dependent.

Obviously it is not all that often that the designer gets a lot of latitude with siting and orientation, especially in urban areas. And the existing building owner has none. It is a lot easier to build high-performance envelope features into the construction of new buildings, than it is to do it after the fact. However, there is an enormous installed base of existing buildings and because the envelope is of such great importance, we must find cost-effective techniques for modifying existing envelopes to significantly increase energy efficiency and reduce environmental impact. Conventional ways to do this involve improving thermal performance by adding insulation, reducing air transfers (infiltration and exfiltration), implementing thermal breaks into building components, increasing the reflectivity or absorption (with vegetation, for example) of surfaces (roofs and walls), and modifying the properties of glazing with films, coatings or exterior shading systems. Another way to do this is collection. *When an envelope surface becomes a collector, it serves two purposes: [1] isolation of the surface from direct energy flow, [2] absorption and conversion of the energy to a useful form.*

If the goal is to leverage the energy assets available at the site, then siting and orientation can play an essential function in accomplishing this goal, even if the designer or building owner has few options. The reason for this is that whatever the siting or orientation of the envelope, harvesting the energy available can be maximized by observing simple principles based on geometry, solar radiation and local weather conditions.

Consider that there is one fundamental building geometry, the rectangle, which can be defined as low-rise (1-story in height), mid-rise (2-4 stories), or high-rise (5+ stories).

All other standard geometries: square, L, T, H, or unusual building geometries: circle, oval, octagon, etc., can be assembled or derived from the rectangle with envelope load factors adjusted for interior sides.

Consider also that there are 8 fundamental orientations: north, northeast, east, southeast, south, southwest, west, northwest as shown in Figure 7-6.

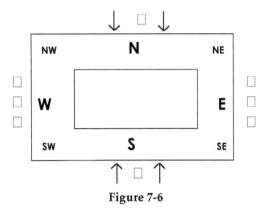

Figure 7-6

Every building geometry and orientation can potentially leverage the energy flows around it, regardless of its siting and local climatic conditions. Wind, rain and solar energy can be predicted for any given latitude and longitude based on average data supplied by NREL or NWS/NOAA (see below) or comparable international agency, and then collected, stored, or transformed for the purposes of supplementing the resource and energy requirements of the building. Every square foot of the envelope can be used to calculate its collection potential for absorbing and translating energy flows. The larger the building, the greater the potential collection area of the envelope. It is practically axiomatic that as the cost of supplying energy remotely with conventional sources increases, the value of collection on sight will rise. Just as grid parity[10] for solar electricity will be reached when photovoltaic panels reach efficiencies of approximately 40% (the relative efficiency of the average coal plant, for example), on-site collection of water and electricity generation from wind & solar energy using envelopes will become cost-effective when it becomes equal to or less expensive than obtaining them remotely.

The updated 1991-2005 National Solar Radiation Database holds solar and meteorological data for 1,454 locations in the United States and its territories. The National Climatic Data Center provides primary distribution of the NSRDB, but this site holds a solar research version of the NSRDB with additional solar fields (without meteorological data). See Figures 7-7a, 7-7b, and 7-7c.

Thus high-performance strategies for the envelope can and should be focused on leveraging and aggregating energy transfers by redirecting flows through collection or diversion to the advantage of the building and its occupants, ideally without disadvantaging other buildings in the external microclimate. That is a tall order, to be sure, but the point of high-performance is to set ambitious goals, and achieve them.

ENERGY TRANSFERS

Principles

Energy transfers through the envelope have to do with flows of heat, light and sound. Air and moisture also flow in and out of the enve-

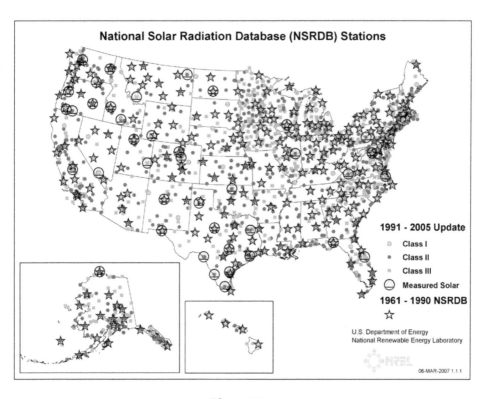

Figure 7-7a

Texas: Full Year 2011 Observed Precipitation
Valid at 1/1/2012 1200 UTC- Created 1/3/12 21:42 UTC

Figure 7-7b

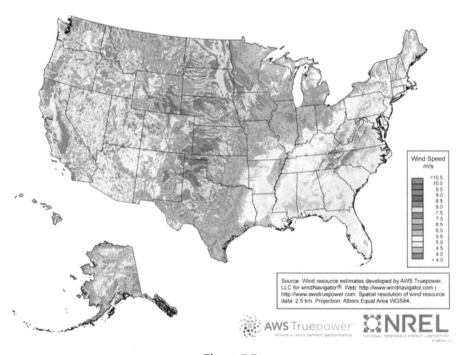

Figure 7-7c

lope largely as a result of heat differentials. The ability of the envelope to control or mitigate these flows has to do with a number of factors, including:

[1] the rate and intensity of the flows (measurable from regional climatic and external microclimatic data),

[2] the orientation of the envelope facades to the respective flows (part of design analysis),

[3] the performance characteristics of the materials in the building components (acquired from lab testing and on-site performance),[11] and

[4] the building usage.

Initially, the analysis of the energy flows related to the envelope can be depicted graphically as shown in Figure 7-8.

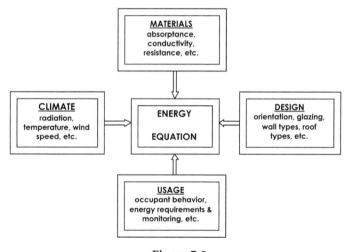

Figure 7-8

To be high-performance, the envelope needs to be able to adapt to a wide range of conditions and this adaptability can be designed into a new or a retrofitted building. A key part of the goals of high-performance buildings when it comes to energy and the envelope is this: the recognition that there is no one material in the envelope which can be optimized to perform at a high level in all orientations, climates and occupancies. Thus materials should ideally be "tuned" for different facades, different flows and different usage patterns.

Material Properties

We've already spoken about the principles of climate and orientation data. Here we can describe the properties of materials[12] which affect the performance of the envelope:

- *absorptance* (ability of a material to absorb radiation)

- *conductivity* (thermal conductance of a material)

- *resistance* (R-value: overall resistance of a material to thermal conductance. U-Value is used to rate the overall resistance of an assembly of materials)

- *emissivity* (ability of a material to emit energy by radiation)

- *reflectance* (the percentage of incident radiation reflected by a material)

- *transmittance* (the percentage of incident radiation which passes through a material)

With respect to materials, the purpose then of high-performance envelope strategies is to engineer and orient component systems *to optimize energy flows to the advantage of the building and the occupants.* When an energy flow contacts the surface of the envelope, the material responds and transfers the energy in accordance with its physical properties. If the energy flow is undesirable, then the envelope materials are usually engineered to try to reflect or resist the flow. But if it reflects most of the energy into the external microclimate, then it may contribute to "heat islands" or cause glare (in the case of glass or metal) on adjacent buildings or public spaces. If it absorbs most of the flow and the undesirable energy has passed through the envelope, then it is inevitably conducted into the interior of the building and must nearly always be mitigated with mechanical equipment. If the energy flow is desirable, then the materials are engineered to absorb, conduct or transmit the flow, yet in a controlled manner. But this can also be problematic, since materials typically do not change their physical properties in accordance with varying flow rates and building orientations.

Obviously then, the overall objective would be to design, engineer and orient envelope materials to maximize efficiency and improve the internal environment for the occupants of the building regardless of location, orientation or flow rates. But are envelopes designed with that goal in mind? As it turns out, the answer the vast majority of the time is no. In the age of inexpensive electricity based on fossil fuels and the use of mechanical equipment and artificial lighting, envelopes are almost always designed for low cost and ease of construction. The problems of heat and light are left up to electrical, mechanical and lighting engineers and shade manufacturers. So typical strategies have focused on:

[1]   increase insulation values by lowering the thermal conductivity of the material,

[2]   decrease air infiltration by making the envelope "tighter,"

[3]   decrease absorption or transmittance by increasing reflectance, and

[4]   increase efficiency of mechanical equipment and electric lighting.

All of these strategies have advantages and disadvantages and are important and part of a good high-performance approach. But all of these strategies, while essential and ongoing, do not include the obvious long-term benefits of *absorbing and translating energy flows* into useful energy for the building through collection.

Energy Usage
Let's say you have a warehouse building 500′ long, 100′ wide and 30′ tall located in Tucson, AZ. It is possible to calculate the average annual quantity of solar energy falling on the east, west and south façades and the roof of the building and thus its capacity as a collector. It turns out that this is a lot of energy over the life of the building. And what percentage of the energy is lost to reflection, absorption and transmission and must be removed by mechanical means from the inside of the building? Almost all of it. The roof alone in this example represents a 50,000-sq.-ft. collection area for solar energy (and rainwater; every drop counts in the desert).

Now consider that in the US alone there are tens of millions of square feet of envelope, walls and roof surfaces, especially in industrial and warehouse settings which are doing little else when it comes to energy except wasting it by reflection, absorption and extraction from the interior with mechanical equipment. This represents potentially vast collection areas for rainwater and solar energy.

Figure 7-9[13]

What if you could efficiently collect and translate most of this wasted energy into a useful form? Geometry plays an important part in this kind of assessment as well as the "energy intensity" of the building. Energy intensity depends on usage and occupancy as well as efficiency. How much energy per square foot does the building use? How much energy per square foot can the building produce on its own? Multi-story Class A office buildings obviously use a lot more energy per square foot than warehouse buildings. But the former, although they may have large collection areas (sq. ft. of envelope), unless they stand alone with few buildings around them, these areas may not be well exposed to wind, water and solar energy. On the other hand, warehouses and industrial buildings, although they use a lot less energy per square foot, they typically have large, fully exposed envelopes (roofs and walls) which can serve as collectors. It is quite possible that these kinds of buildings could become *net-energy producers* due to their low energy usage.

The following are useful concepts when it comes to energy use and collection which can inform the process.

Energy Intensity

"Energy intensity is measured by the quantity of energy required per unit output or activity, so that using less energy to produce a product reduces the intensity."[14]

High-performance envelopes conserve energy and thus reduce intensity.

Energy Density

"Energy density is the amount of energy stored in a given system or region of space per unit volume."[15]

An existing building has a large amount of energy embedded in it and therefore high-performance building strategies should emphasize retrofit, reuse and repurposing.

Energy Intensity of a Building (EIB)[16]

This is the energy consumption (type of energy) per unit area, for example: *kWh/square foot.*

A high-performance envelope will reduce the energy consumed per unit area and thus the energy intensity of the building.

Energy Use Intensity: EUI (from IgCC)

EUI = TAE/SF (where TAE = total annual energy projected to be consumed on site, including renewable energy; and SF = building gross square footage) (Section 602.2.4.1), IgCC Public Version Synopsis, p. 20.

A high-performance envelope will reduce the total amount of energy consumed and thus the energy use intensity of the building.

For our own purposes with HP envelopes we can expand and modify the above concepts to develop the following:

Energy Production of an Envelope(EPE)
The energy which can be generated (collected and converted) per unit area of an envelope, for example: *kWh/square foot of wall or roof.* (This concept could be broadened, of course, to include the energy production of a building site which would include the envelope of the building and any land also available for collection and distribution purposes.)

Conversion Efficiency of the Envelope(CEE)
This is the ratio of the energy absorbed to the energy generated per unit area.

Obviously then the EPE depends on the CEE, and we can develop another potentially useful ratio, the *energy generation coefficient of the envelope (EGCE),* given by: *EPE/EIB* = the ratio of the energy potentially generated by the envelope per unit area to the energy consumed by the building per unit area

These kinds of ratios could help guide us in determining appropriate strategies for constructing high-performance envelopes with various building types in different locations. It could also help us to determine a sensible mix of component modifications to improve thermal efficiency along with collection techniques to generate energy, adjusted of course, for cost considerations. Since envelope component modifications and energy collection systems are relatively expensive per unit area, it is important that we have tools to analyze long-term performance in a wide variety of conditions.

HEAT[17]

The building envelope is a thermodynamic system, and the integrated, high-performance approach recognizes this central fact. All the parts of the system should be considered to get a comprehensive analysis and obtain coherent outcomes. Leaving aside external and internal climatic conditions, heat transfer through the walls and the roof of the envelope, either in or out, is governed by the conductance, absorption, reflectance and resistance properties of the building materials which make up the component systems. The basic strategy is to try to slow down or block heat transfers using insulating materials, air spaces and reflective coatings or metallic films. This typically utilizes insulation materials that tend to inhibit thermal conductance using fibers or cellular foams with interstitial air spaces. Reflective coatings are also applied to the components to mitigate energy transfers with radiant barriers although these effects are limited to emissivity. When the R-values of all of the wall component materials are summed, one can obtain a total resistance value for a wall (or roof) assembly.

The multiple R-values are built up from assemblies whose components include [1] insulation, [2] wall panels, [3] roofing, [4] glazing. Air infiltration, the leakage of air into the building, and air exfiltration, the leakage of air out of the building, can be major contributors to unwanted heat transfers. High-performance strategies can also involve secondary collection systems, such as "green" roofs and walls and building integrated renewable energy generation.

ROOF SYSTEMS

With the exception of buildings over three stories in height, the roof portion of the envelope typically makes up a large percentage of the total envelope exposed to external climate, and so it is not just a waterproof surface that diverts water, it is as an absolutely essential focus of the high-performance envelope. Consider a 1-story commercial building, 50' (east-west exposure) x 150' (north-south exposure) x 20' high. The total square footage of the roof is 7500, which is subject to direct solar radiation almost all day long. The total square footage of the walls is 8000. If we leave out the north wall of the building, that leaves 5000 sq. ft. of envelope wall exposed to direct solar radiation. So the roof in this example is 48% of the total envelope or 60% of the envelope exposed to direct solar radiation. Also, heat rises and it wants to escape through the roof in cold temperatures. Since the roof area is exposed almost all day to direct solar radiation, this also makes it a heat sink, which can reradiate energy into the building or back into the external microclimate.

Reflective
High-performance commercial roofing systems typically combine panelized foam insulation with reflective roof membranes and coatings.

The goal is to reduce conductance, in or out, by increasing R-value at the deck and reduce absorption and heat transfer in from the exterior, by reflection with the membrane surface or coating (usually white or aluminized). The system must also, of course, be designed to control vapor transmission in the components as well. Some foam panels may also have reflective films on one or both sides to increase the overall reflectivity of the system.

There a number of foams and other materials available for roofing insulation, but the high-performance material of choice has certainly become the closed-cell foam polyisocyanurate (ISO). Compared with its close cousin, polyurethane (PUR), ISO is denser, stiffer, has a bond breakdown temperature of 200°C (increasing its life through heat cycling), and has an R-value of 5.6 per inch.[18] ISO also happens to have a lower environmental impact than other foams and is to some extent recyclable. Figure 7-10 shows a mockup of an ISO panel from XtraTherm[19] (also shown as vertical wall panel) with a steel deck and capsheet.

Figure 7-10

High-performance roofing materials can be single-ply membranes such as TPO (thermoplastic olefin), EPDM or PVC. White reflective roof membranes coupled with high R-value foam insulation panels are estimated to reduce energy costs by anywhere from 10-15% by cutting heat transfer through the deck and reducing air-conditioning loads. Developments in SBS (styrene butadiene styrene) modified systems have added "cool roofing" granules to the cap membrane to produce results similar to white reflective single-plys. The Cool Roof Rating Council[20] and ANSI produced a joint standard in 2010[21] which provides methods for rating the solar reflectance (SR) and thermal emittance (TE) of roofing products, also governed by ASTM E 903.[22] Typical values for SR for these types of products are high, anywhere from .7 to .8, with 1 equaling total reflectance. Roofing which reflects heat rather than absorbing it, not only cuts loads in the specific building application, it also cuts some of the "heat island" effects in dense areas which contribute to increased air-conditioning loads on adjacent buildings in the surrounding external microclimate.

When installing a new roof is not applicable, high-performance reflective roof coatings can, in many cases, be applied over an existing substrate to substantially increase the SR and the TE of the existing roof profile. These coatings are either white or aluminized to increase reflectance and can obviously at the same time improve waterproofing. It is not feasible to make general statements about overall savings using reflective roof coatings as there is so much variation from one existing building to another, however, the Roof Coating Manufacturers Association[23] and the Reflective Roof Coatings Institute[24] provide guidelines for calculating energy savings in various scenarios. The DOE has a fairly simple Cool Roof Calculator[25] that can help do a preliminary appraisal for either new high-performance roofing systems or applied coatings. A more in-depth analysis can be done with the Roof Savings Calculator from LBNL and the Oak Ridge National Laboratory.[26] Building owners and managers may also want to perform life-cycle cost analyses on different high-performance roof system scenarios and these can be conducted using ASTM E917-05(2010), Standard Practice for Measuring Life-Cycle Costs of Buildings and Building Systems.[27]

Absorptive—Green Roofs

As opposed to roofing which is reflective, "green roofs" actually absorb heat instead of reflecting it back into the external microclimate or transferring it into the building. The vegetation in the green roofing system effectively translates the solar radiation into energy for growth. The green roof follows the high-performance principle of collection, as the energy is absorbed and translated into a useful form. A green roof can cut heat transfer through the roof by at least 50%. Green roofs also do a better job of mitigating the "heat island" effect than reflective roofing, since they reradiate little or no heat into the environment. Since green roofs absorb rainwater, they can also contribute to stormwater retention. A typical layout for a green roof system is shown in Figure 7-11.[28]

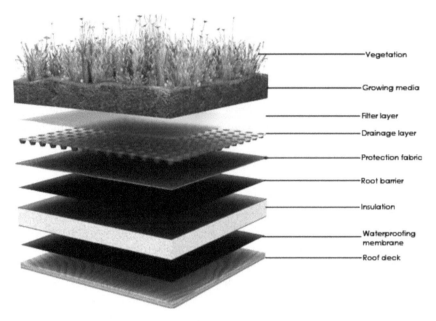

Vegetation
Growing media
Filter layer
Drainage layer
Protection fabric
Root barrier
Insulation
Waterproofing membrane
Roof deck

Figure 7-11

Green roofs require special preparation in the substrate for moisture penetration issues and the roof structure, since weight is an issue. The weight of a green roof can vary from 18 to 30 pounds per square foot, not including water absorption, which may impose limitations on existing buildings. Nevertheless, they are sometimes a good choice with older, traditional buildings which have very substantial concrete or masonry roof systems which can carry the accompanying structural loads. The ASTM publishes two standards regarding green roofs:

- ASTM E2400—06 Standard Guide for Selection, Installation, and Maintenance of Plants for Green Roof Systems

- ASTM E2397—11 Standard Practice for Determination of Dead Loads and Live Loads Associated with Vegetative (Green) Roof Systems[29]

The most widely accepted, oldest and most complete standard in the world is the German FLL: Standard "Guidelines for the planning, execution and upkeep of Green Roof sites" and are from The Landscaping and Landscape Development Research Society (FLL) in Bonn, Germany. In 2009, the International Green Roofing Association (IGRA)[30] published its Proceedings of the International Green Roof Congress: Green Roofs-Bringing Nature Back to Town. This document synthesizes the German Standard and makes it available for English-speaking readers.

One of the most famous installations of a green roof is on City Hall in Chicago (Figure 7-12).[31]

Figure 7-12

In the fall of 2012 I had the opportunity to visit the Greenpoint Manufacturing and Design Center (GMDC) in Brooklyn, NY. The GMDC[32] is a non-profit real estate development corporation which has been in existence for about 20 years. It takes old buildings which are abandoned or underutilized and converts them into lease space for boutique manufacturers and artisans. Because it is non-profit, it can lease at below market rates and it keeps its buildings 95% leased at all times. GMDC makes a point of integrating energy efficiency measures into its retrofits wherever possible including the use of distributed on-site power. Although it does not, strictly speaking, have a green roof, one of their buildings at 810 Humboldt Street[33] has a very clever combination of high-performance roof attributes. It has a reflective roof system with a successful, for-profit greenhouse facility which occupies approximately 50% of the roof surface, and the rest of the roof area is occupied by a 60kW photovoltaic array. So the roof is high-performance, it collects solar energy and converts it into electricity to help power the building, and it provides space for growing plants where land is at a premium. We call that aggregating utilization factors, a fundamental high-performance strategy. (See Figure 7-13.)

Figure 7-13

WALL SYSTEMS

High-performance wall systems can be thickwall or thinwall, but except in older buildings or unusual circumstances, they are primarily going to be thinwall. Retrofits to older buildings can involve adding high-performance components to existing envelope but the principles of the new components remain the same. As far as heat transfer is concerned, since the north side of a building rarely receives direct solar radiation, heat transfer is generally out of the building except for the effect of ambient air in the cooling months. In general, the south side of the building receives the most direct solar radiation, followed by the west and the east. In the heating months then it may be advantageous to absorb solar radiation and transfer it; in the cooling season, reflect it, or collect it and convert it. The point is that, regardless of latitude or solar exposure, different orientations would ideally call for different strategies with high-performance wall systems that maximize and leverage the energy assets available whenever possible.

Currently the technology for HP wall systems includes conventional stud-framed, panelized components, masonry, composites and integrated material systems which perform similar functions to HP roofing systems, reduce conduction with insulation, provide thermal bridging[34] with component isolation, increase reflectance with films and coatings, control vapor transmission, and reduce air infiltration with sealants and tapes. There are literally hundreds of innovative approaches to HP wall systems, residential and commercial, many of which are still under development and evaluation. We'll look at some of the major approaches.

Conventional

Conventional wall systems can be defined as stud-framed with siding or masonry veneer. They all contain similar components and are characterized by multiple construction assembly operations by different trades—wood or metal framing members, insulation materials, sheathing, vapor barriers, sealants and tapes, composite, metal, wood sidings or masonry veneers. There have been concerted efforts over the past ten years to make these types of systems more high-performance by increasing insulation capacity with additional wall-thickness (2x6 studs instead of 2x4), air spaces, adding high R-value structural foam sheathing panels, carefully taping and sealing joints, fasteners and holes, foam insulating walls in place, and mitigating or eliminating thermal bridging. Nevertheless, conventional wall systems, although greatly improved from the standpoint of energy transfer, are still vulnerable to the large number of different components involved and the potential for the uneven quality of assembly. Figure 7-14 outlines a standard residential wall with some high-performance measures.[35] Figure 7-15 depicts a commercial wall section using Owens-Corning's Foamular XPS insulation panel.[36]

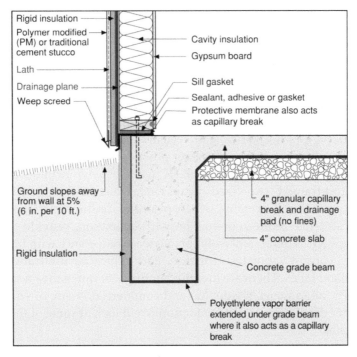

Rigid insulation

Polymer modified (PM) or traditional cement stucco

Lath

Drainage plane

Weep screed

Cavity insulation

Gypsum board

Sill gasket

Sealant, adhesive or gasket

Protective membrane also acts as capillary break

Ground slopes away from wall at 5% (6 in. per 10 ft.)

4" granular capillary break and drainage pad (no fines)

4" concrete slab

Rigid insulation

Concrete grade beam

Polyethylene vapor barrier extended under grade beam where it also acts as a capillary break

Figure 7-14

Using a variety of these components and fairly simple techniques have increased average wall insulation values in conventional wall systems from about R-13 to as much as R-26. Tightening up the envelope with close attention to preventing air leaks though components has also decreased air infiltration considerably. [37] See Figure 7-16.

Composite Panel

Because conventional wall systems involve so many different assembly operations and are more prone to breaks in the system, the goal of producing tighter, integrated wall systems with fewer parts, more pre-manufactured connection procedures and less on-site assembly has rapidly advanced the development of composite panel systems. Pre-manufactured systems are not necessarily less expensive overall, but quality can be more uniform and performance more easily predicted since complete systems are tested in advance and installation procedures are more easily controlled. Some of the systems also have struc-

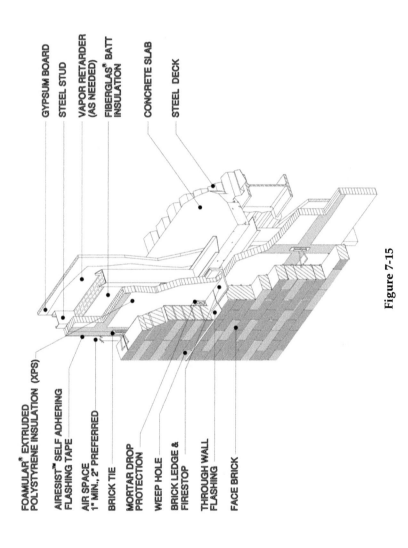

GYPSUM BOARD

STEEL STUD

VAPOR RETARDER
(AS NEEDED)

FIBERGLAS® BATT
INSULATION

CONCRETE SLAB

STEEL DECK

FOAMULAR® EXTRUDED
POLYSTYRENE INSULATION (XPS)

AIRESIST™ SELF ADHERING
FLASHING TAPE

AIR SPACE
1" MIN., 2" PREFERRED

BRICK TIE

MORTAR DROP
PROTECTION

WEEP HOLE

BRICK LEDGE &
FIRESTOP

THROUGH WALL
FLASHING

FACE BRICK

Figure 7-15

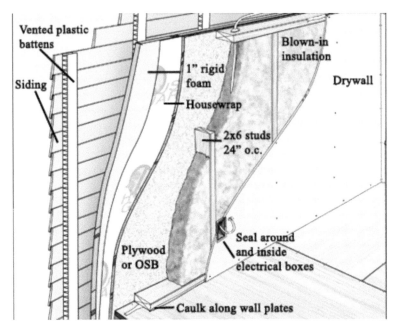

Figure 7-16

tural properties which make them stand-alone.

Composite panel systems nearly all use foam as the primary insulating material, and can combine metals, plastics and wood. There has been a drive as well towards producing systems which are not only high-performance in terms of energy but also utilize recycled materials and employ manufacturing techniques which are environmentally benign. A system from Centria is modeled in Figure 7-17.[38]

It is easy to see that combining an R-14 or R-21 metal wall panel system with a stud wall cavity using R-13 batt insulation or R-15 foamed in place insulation could result in total insulation for this type of wall system approaching R-30 to R-40 with a very tight envelope. Centria's Formawall system has also earned the Silver Certification for sustainable manufacturing processes from the Cradle to Cradle Institute.[39] Allowing for vapor transmission considerations, these kinds of systems could also find application in retrofit situations.

Another composite wall panel system which has been used primarily in residential and light commercial construction for many years is the Structural Insulated Panel System (SIPS). The panels are composed of Expanded (EPS)or Extruded (XPS) Polystyrene Foam laminat-

Figure 7-17

ed between sheets of Oriented Strand Board (OSB). The system comes to the construction site in large sections which can be assembled rapidly. The Structural Insulated Panel Association publishes technical reports and general information on the products. For example, a 6-1/2" EPS panel is rated at R-21 and an XPS panel at R-30[40], the latter providing almost 50% better insulation value than a 2x6 conventional wall system with fiberglass batting. The SIPS system also provides 2-3 times the air infiltration performance of any conventional wall system.[41] See Figures 7-18 and 7-19.

Green Walls

The green wall system either involves vegetation growing directly on the surface of the building or is more effective and easier to maintain if it employs a stainless steel cable or trellis system to provide a platform for vines or similar growth to shield the envelope. This approach works on the idea of shading the wall of a building from direct solar radiation in the cooling season and, if the vegetation is deciduous, then it loses its leaves in the heating season and allows absorption of solar radiation from collection. The green wall can also provide some soundproofing. Although it is difficult to measure the impact of this strategy from a testing standpoint, it is worth mention-

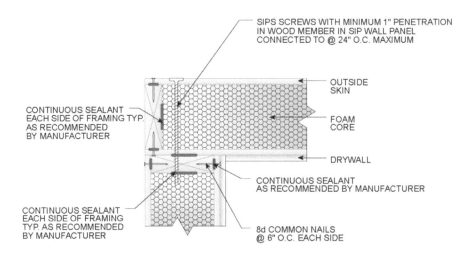

SIPS SCREWS WITH MINIMUM 1" PENETRATION IN WOOD MEMBER IN SIP WALL PANEL CONNECTED TO @ 24" O.C. MAXIMUM

OUTSIDE SKIN

CONTINUOUS SEALANT EACH SIDE OF FRAMING TYP. AS RECOMMENDED BY MANUFACTURER

FOAM CORE

DRYWALL

CONTINUOUS SEALANT AS RECOMMENDED BY MANUFACTURER

CONTINUOUS SEALANT EACH SIDE OF FRAMING TYP. AS RECOMMENDED BY MANUFACTURER

8d COMMON NAILS @ 6" O.C. EACH SIDE

WALL-TO-WALL PANEL CONNECTIONS

CORNER WALL CONNECTION

Figure 7-18

Figure 7-19[42]
Little Bighorn Health & Wellness Center

ing as a high-performance option which can have similar effects to the green roof. It can be used in hot climates to significantly reduce heat transfer by shading. See Figure 7-20.

Masonry

The insulating concrete form (ICF), used for residential and light commercial construction, has been in use in the US since the 1970s. It is recognized by nearly all code jurisdictions and is a well-established and thoroughly tested system. The ICF block is typically made of polystyrene or polyurethane foam and is hollow to allow the introduction of rebar and poured concrete, forming a structurally sound wall with outstanding insulating properties, soundproofing, and resistance to fire and violent weather. R-values for an ICF wall system can vary from R-21 all the way up to and exceeding R-50 and the system has near zero thermal bridging. The wall section shown in Figure 7-21 is from Build-Block.[44] See also Figure 7-22.

Figure 7-20[43]
LEFT: MMA Architectural Systems Ltd. RIGHT: Carl Stahl

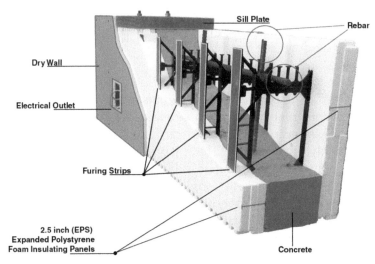

Figure 7-21

Figure 7-22[45]
LEFT: Waterside Condominiums, Fort Myers, FL, in *Structure Magazine*
RIGHT: 4000 sq. ft. Single-Family Home by Midwest Modern, LLC

Collector

One of the principles we have emphasized repeatedly in high-performance buildings is the idea of collecting energy, especially with the exterior of the envelope, rather than reflecting it into the microclimate or absorbing it and reradiating it into the structure (although that could be advantageous in cold climates). In passive solar architecture it is part of the design process to leverage southern orientations to absorb solar radiation as heat and use it in the interior space. There have been numerous strategies employed with varied success to heat water, air, etc. in the envelope wall system and circulate those fluids into the building systems. The problem there has been to justify the additional cost in the construction of the envelope and the inevitable maintenance problems that come along with it. But in high-performance wall systems, ultimately we need systems which can absorb and translate solar energy wherever it falls, ideally convert it into electricity and integrate that *as part of* the exterior of the envelope wall.

Building integrated photovoltaics (BIPV) is one of the more promising approaches to this goal and as the efficiency of PV increases the application will become more practical. The idea in new buildings is to build wall panels with BIPV that replace existing wall panel systems so that PV is not a system attached on top of existing substrates. As a report from NREL states: "BIPV can be directly substituted for other cladding materials, at a lower material cost than the stone and metal it replaces."[46] BIPV has been manufactured as composite wall panels, roof shingles, spandrel panels, etc. BIPV do generate heat in the collection process like all PV panel systems which must be mitigated, but systems have been developed to handle this.[47] There have also been systems created to combine electrical production and solar thermal concurrently.[48] BIPV applications are still in the development stage but the market is increasing steadily for this solution. "The global BIPV market is expected to grow from $4.33 billion in 2009 to $12.73 billion in 2016 at a CAGR of 16.9% from 2011 to 2016."[49] See Figures 7-23 and 7-24.

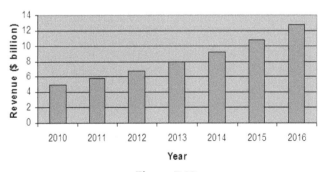

BIPV Market ($ billion)

Figure 7-23

Figure 7-24[50]
LEFT: CIS Tower, Manchester, England
RIGHT: Solar Roof Tile

An in-wall approach is shown in Figure 7-25, and another approach is to use PV panels as a standoff from an existing wall, providing shading, envelope isolation and power production at the same time, as shown in Figure 7-26. This technique allows air circulation between the PV system and the envelope and mitigates heat build-up.[51]

Retrofit

There are tens of thousands of existing commercial buildings in the US which have low-performing envelopes with high thermal transfer, air infiltration and possibly vapor transmission issues. High-performance wall systems are obviously most easily designed for new buildings but some creative solutions are being developed to retrofit existing buildings with high R-value, low air infiltration wall systems. Filling wall cavities with insulating materials is often impractical and because of the logistics of attaching materials to existing walls, either internally or externally, high-performance wall components which are added on must generally be lightweight, thin, and cannot compromise the vapor transmission (moisture balance) requirements of the entire system.[52] The principle of an exterior retrofit is well illustrated here with this wall section, a "ven-

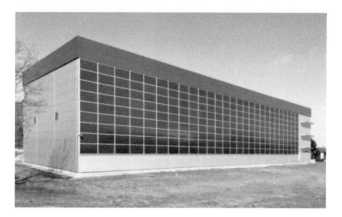

Figure 7-25

Figure 7-26

tilated façade" from KraSpan.[53] The space between the composite wall system and the insulation layer is essential as it allows a "drainage plane" for moisture control and air circulation (see Figure 7-27).

There are numerous ways to do it, but one recent example we can look at is very instructive and addresses some of the critical issues. Obviously existing buildings with serious heat transfer problems are the best candidates and this has led to a lot of research into retrofits on older masonry buildings in cold climates which have very low R-values. Concrete walls have similar issues. But the principles are also applicable to existing buildings with low R-values in hot climates; losing cooled air is just as wasteful and expensive as losing heated air.

The Castle Square Apartments in Boston were proposed for a "deep energy retrofit" in 2011. In addition to air-sealing, new windows and a reflective roof system, this project was to include the addition of a high-performance exterior wall system retrofit.

Figure 7-27

"The mid-rise building was originally constructed with no wall insulation, a design that yielded an abysmal R-3 rating. The retrofit will increase that value to R-40...."[54] The general approach is as shown in Figure 7-28; analysis done by Building Science Corporation.[55]

High-performance wall retrofits can be costly but the long-term improvements in comfort, energy efficiency and O&M costs have to

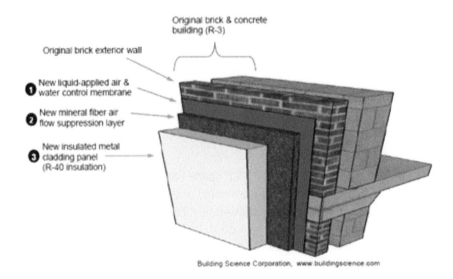

Original brick & concrete building (R-3)

Original brick exterior wall

❶ New liquid-applied air & water control membrane

❷ New mineral fiber air flow suppression layer

❸ New insulated metal cladding panel (R-40 insulation)

Building Science Corporation, www.buildingscience.com

Figure 7-28

be considered in the assessment. HP Envelope improvements do not become relatively less efficient over time or degrade like mechanical or electrical equipment and this should be part of the life-cycle cost analysis of any major energy retrofit.

Figure 7-29. BEFORE retrofit

Figure 7-30. AFTER retrofit

FOUNDATION & FLOORS

The principles of managing heat flows through foundations and floors(ground level) are very similar to those employed with wall systems, with the applications considerably less varied. Obviously with a few exceptions high-performance measures must be designed in and implemented with new buildings; it is difficult to retrofit high-performance into an existing foundation. In climates with a lot of solar radiation, it is possible to install a water piping system in the concrete which can circulate water heated by solar thermal collectors on the roof to heat the floor. It is not out of the question to insulate on top of an existing concrete slab floor with high-density ISO foam panels before applying a finish flooring system to improve R-value. And wood floor framing systems in existing buildings over a crawlspace can also benefit by having fiberglass batting or sprayed foam installed between the floor joists.

High R-value foam insulating panels and applied coatings for vapor barriers are used to isolate concrete foundation structures from ground contact and resist heat transfers out and cold transfers in. Since the ground under a concrete floor maintains a more constant temperature, thermal bridging occurs typically in perimeter conditions where the temperature differential is higher and other building components come into play.

Figure 7-31 shows a few basic approaches from the DOE's Energy Efficiency and Renewable Energy site.[57]

Notice that in the diagram to the far left, the insulation board extends up past the intersection of the bottom plate of the wall and the foundation beam and past the intersection of the plate and the stud. This is to counteract the heat flow that occurs through the slab and wall components at the perimeter. Figure 7-32 profiles a section of a concrete floor using an under-slab XPS sheet from Expol.[58]

Figure 7-33 is a diagram from DOE's Buildings Foundation Book section 3.1 showing a strategy for insulating a floor in a vented crawlspace with batting and ISO panels.[59]

A very useful site for residential construction is LowEnergyHouse. com, which provides the detail in Figure 7-34 for a high-performance slab floor layout.[60]

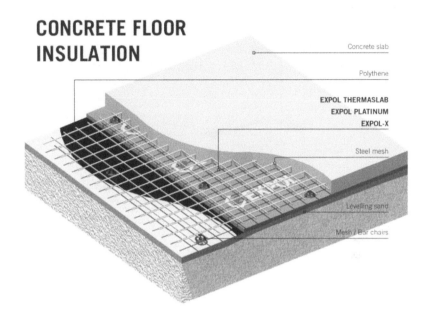

Figure 7-31

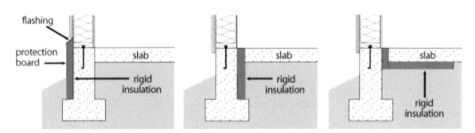

Figure 7-32

Vapor Transmission (VT), Air Infiltration (AI), Indoor Air Quality (IAQ)

The proper design of a high-performance wall system with regard to its impact on vapor transmission, air infiltration and indoor air quality is an essential part of the equation. New buildings will factor in all of these issues at the design and engineering stage and assess them along with testing data from the manufacturers of the components. Careful control and inspection during the construction process is required to insure consistent results.[61] "Tighter" envelopes which highly restrict AI

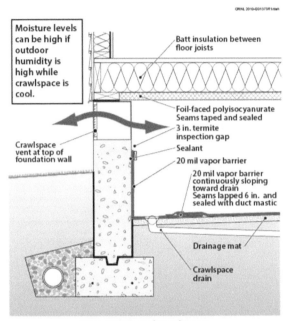

Figure 3-9: Vented Crawl Space
with Insulation in the Ceiling

Figure 7-33

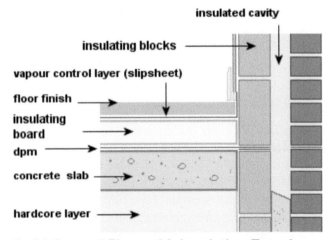

Solid Ground Floor with Insulating Board

Figure 7-34

can have serious effects on air-exchanges and IAQ, and envelopes with multiple wall components having impermeable, semi-permeable and permeable layers present challenges for VT and moisture control. These issues are governed by numerous standards from ASTM and ASHRAE:

ASTM Standards on Indoor Air Quality: 4th Edition. This includes:
- D22.05 Indoor Air
- E06 Performance of Buildings
- E30 Forensic Sciences
- E35 Pesticides, Antimicrobials, and Alternative Control Agents
- E47 Biological Effects and Environmental Fate

ASHRAE: The Indoor Air Quality Guide: Best Practices for Design, Construction and Commissioning,

ASTM E96/E96M—12 Standard Test Methods for Water Vapor Transmission of Materials.

Whenever modifications to the envelope of an existing building are contemplated, a careful assessment of existing conditions should be done first, including the structure and functioning of wall and roof components. It is estimated that anywhere from 25-40% of energy losses can be due to air infiltration. Therefore testing to detect air leakage and heat transfers is essential to evaluating strategies for improvement, and these procedures can include infrared thermography, smoke tracer, anemometer, sound detection, bubble detection, and tracer gas techniques.[62] Existing wall systems may already have problems with VT and moisture control and adding layers to the system can make matters worse or create new moisture control problems. Air infiltration can be modeled using Energy Plus[63] and standard methods have been developed to evaluate existing conditions:

ASTM E1186—03(2009) Standard Practices for Air Leakage Site Detection in Building Envelopes and Air Barrier Systems,

ISO Standard 6781 Thermal Insulation—Qualitative Detection of Thermal Irregularities in Building Envelopes—Infrared Method,

Moisture Analysis and Condensation Control in Building Envelopes, ASTM, 2001

LIGHT

The fundamental importance of natural light[64] to the health and well-being of building occupants has been known since ancient times. It is arguably the most important part of the functioning of the envelope. The design and application of daylighting in modern commercial buildings is quite involved and requires a skilled combination of artistic and scientific principles. Dozens of studies have been conducted in the US alone since the 1940s describing the positive and essential effects natural light has on increasing the productivity and decreasing the absenteeism of employees. Many of these studies and seminal research on the subject have been done under the auspices of the Illuminating Engineering Society[65] and the Lawrence Berkeley National Laboratory.[66] A tremendous amount of research and testing into the performance of glass, coatings and shading systems has been done in the last 20 years by the National Fenestration Rating Council[67] (NFRC) as well. If a building does not have adequate and well-designed daylighting, it certainly cannot be called high-performance.

The basic principles of daylighting applied to the structure of the building have been known for a long time. But the technologies to control daylighting with glazing, not so long, and they are still under development. To state it simply: *the objective is to allow enough visible light in to illuminate the space and the tasks to be performed while controlling glare and heat transfers at the same time.* Easy to state. Hard to do. An in-depth treatment of daylighting strategies, as there are a great many, are described in great detail in other volumes.[68] So we show here a few approaches which demonstrate the principles, some of which have been revived from older techniques with some new technologies and which apply to high-performance goals. We will proceed in order of most to least important.

Architectural

From the architectural standpoint there are, of course, only two ways to do it: *side lighting* (fenestration—operable windows, vertical and sloped fixed glazing,) and *top lighting* (skylights, monitors, atria, clerestories, etc.), with numerous variations inside each category.

The optimal design and orientation of a building for high-performance daylight utilization is governed by the following:

[1]   orient the length of the building along the east-west axis
[2]   maximize the north and south facades
[3]   minimize the east and west facades (60′ floor plate depth maxi-
      mum)
[4]   reflected light is optimal rather than direct sunlight

Figure 7-35 is an excellent example of these factors in practice, design by Mithun Architects.[69] Figure 7-36 is a diagram of window and light shelf effects on a south façade from WBDG.[70]

North light is reflected light and is optimal. South light causes glare and unwanted heat gain if it is direct, but when reflected or shaded properly is also beneficial. High-performance design strategies then focus on using architectural features in the building to channel the admission of light and absorption of solar energy, rather than relying on mechanical equipment or interior window shades to mitigate unwanted effects. The key point is that daylighting strategies are best applied in the design of the building, but some of them (skylights for example and in some cases light shelves) can be implemented as well in retrofits.

Figure 7-37 shows some examples of basic top lighting strategies.[71]

Clerestories, monitors and saw tooth structures provide reflected light the majority of the time with skylights usually requiring diffused glazing or shading controls to avoid direct solar gain (unless they are very high off the floor surface). Top lighting with skylights can provide more illumination per square foot than side lighting but is obviously not feasible on multistory buildings without the construction of an atrium. Some examples of the art and mastery of top lighting in the atrium are shown in Figure 7-38.

Bell Labs, a series of buildings which quite probably produced more inventions per square foot than any buildings since Leonardo DaVinci's studios, is a great testament to the importance of design and natural light to productivity (see Figure 7-39).

From the standpoint of energy savings, a study was conducted in 2009 by the Department of Energy on top lighting:

Studies have repeatedly found that daylighting has the potential to real-
ize very large reductions in lighting energy consumption. For example,
the TIAX Controls and Diagnostics Report found that dimming electric
lights in daylit spaces could reduce annual lighting energy consumption
in existing commercial buildings by 40-60% (New Buildings 2001, New

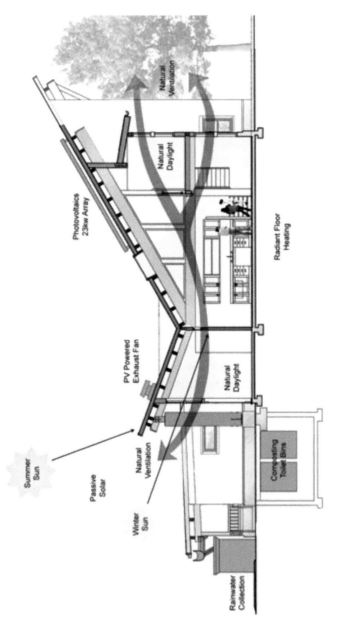

Figure 7-35

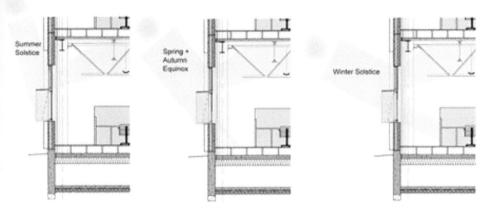

Figure 7-36

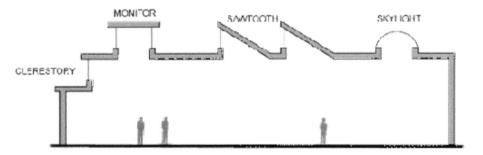

Figure 7-37

Figure 7-38. LEFT: The Guggenheim Museum, NY[72]
RIGHT: The Galleria—Milan[73]

Figure 7-39

Buildings, 2003). Daylighting can be achieved through side lighting (windows) or top lighting (skylights).[74]

Glazing

Determining the solar-optical properties of glazing materials through testing and simulation is no simple task. ASHRAE, ASTM, LBNL, NFRC and others have worked for decades to develop testing and simulation procedures which are fair, accurate, comprehensive, and most importantly, can realistically predict performance in the field. The physics of solar radiation is well understood, but how materials behave under a very wide range of conditions is complicated and sometimes not entirely predictable. Weather conditions, solar irradiance[75], the solar profile angle[76], the behavior of materials under different loads, etc. make prediction a challenging process. We know from experience that glazing, shading and films behave differently based on the solar profile angle. I was personally involved in an on-site testing procedure with a Fellow of the Illuminating Engineering Society who had been in the business for more than 40 years and was a founding member of the NFRC. We conducted a test of the originally installed single ¼" bronze glass, an exterior shade application and a film application on an existing building.[77] The results showed the high effectiveness of exterior shading (solar heat gain reduction in excess of 60%), but also that the glass and the film behaved differently than predicted results based on specifications. Testing methods and simulation programs have improved significantly since that time, but demonstrations in the field are still a very wise idea.

A single sheet of clear glass allows the maximum view and visible light transmission of all glasses. However, unless it is guarded by overhangs or it is on the north side of the building, it also allows the most glare and heat gain as well (heat loss is also a serious problem no matter what orientation). And glass, unlike almost any other building material employed in the envelope is a very poor insulator. To address these problems, glass manufacturers began in the 1940s to develop "solar-control" glass to mitigate those effects. Since buildings were becoming "curtain-wall" structures, the goal, although continually elusive, was to develop a glazing system that allows maximum view, abundant visible light, controls heat gain (and loss) and does not require overhangs or exterior shading to perform. Early efforts involved tinting the glass[78] to make it heat-absorbing and reduce the amount of heat entering the building. The problem was that if it was tinted enough to cut the heat gain, it also cut the visible light transmission and made the outside view look dark. It also did not stop glare or cut heat loss. Therefore, the next step was to produce double-pane, insulated glass to help control heat transfers and improve the U-factor of the glazing system.[79] This was moderately successful in combination with tinted glass to help control heat gain and loss, but overall performance, although better, was still not satisfactory.

This brings us to the manufacturing process chemical vapor deposition[80] (CVD) and the development of "high-performance" or "spectrally selective" glass over the last 30 years. See Figure 7-40.

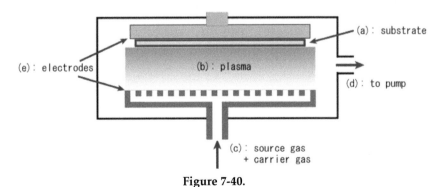

Figure 7-40.

CVD was initially employed in semiconductor manufacturing and is a process which, in the case of flat glass for buildings, involves the deposition of thin films of metal-oxide coatings, having a thickness in nanometers, onto the surface of the glass (substrate). These thin films, also known as Lo-E (low emissivity) coatings, can be designed to "select" wavelengths (visible and infrared), either for reflection or transmission. Since the goals are high visible light transmission, low direct and indirect solar heat gain and loss, and higher insulation value, Lo-E coated glass is combined in an insulated unit to optimize the properties of the coatings to achieve those goals. When these properties are aggregated, Lo-E coatings can help reduce solar heat gain through the window, and also retard heat loss out of the window. The Efficient Windows Collaborative,[81] combining the Center for Sustainable Building Research[82], the Alliance to Save Energy[83] and the LBNL, has developed a very useful on-line tool which can help researchers, designers and building owners to make informed decisions about the use of windows and glazing in any type of building. It profiles dozens of glass types in a variety of combinations with frame properties to determine overall performance of the window/glazing system. LBNL also publishes a software program, COMFEN[84], which integrates LBNL's optimization engine with Energy Plus to guide designers and engineers in the process of calculating loads and lighting requirements, energy savings and help predict future performance.

A simple example (Figure 7-41) can help demonstrate the improvement from clear glass to Lo-E glazing.

The result is a more than 4-fold improvement in the U-factor of the overall system. Recent research and development done by PPG, similar to the example above right, on their triple-silver-coated glass product, Solarban 70XL[87], has brought the development of "spectrally selective" glass closer to the theoretical limit obtainable with the materials and the technology. James J. Finley, PhD of PPG says "manufacturers are approaching what is considered to be the physical limit of thin-film-coated spectrally selective glazing—a light-to-solar-gain (LSG) ratio[88] of 2.5—offering unprecedented levels of daylight with minimal solar heat gain."[89] Solarban 70XL promises 64% visual light transmittance and a solar heat gain (SHG) coefficient of 0.27, meaning it blocks up to 73% of the sun's energy. "We're not going to get much better performance than that," says Finley.

Still, it is important to note that as impressive as these devel-

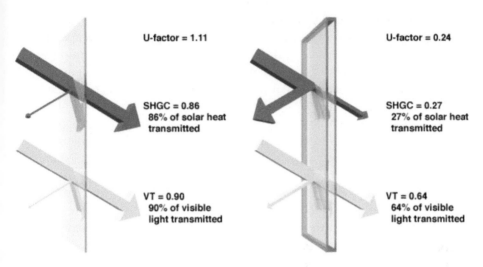

U-factor = 1.11

SHGC = 0.86
86% of solar heat
transmitted

VT = 0.90
90% of visible
light transmitted

U-factor = 0.24

SHGC = 0.27
27% of solar heat
transmitted

VT = 0.64
64% of visible
light transmitted

Figure 7-41.
LEFT: Single Clear Glass[85]
RIGHT: Lo-E Insulated Glass with Argon-Gas fill[86]

opments are in high-performance glass, they do not overcome the problems of glare or heat transfer. To follow through, U-factor is used primarily to rate the overall thermal performance of windows and glazing, and is the reciprocal of R-value, which is used to rate the overall performance of walls and roofs. Therefore the R-value of the highest rated glazing system available as shown above is $1/U-.24 = R-4.16$. The R-value of a typical 2x4 stud wall with R-11 fiberglass batting in it and sheathing and drywall attached is about R-14, almost 3.5 times the value of the glass. Clearly there is still a lot of heat transfer through the glass, despite the impressive technological advancements. In other words, spectrally selective glass is a huge leap forward but does not eliminate the need for glare control under direct solar radiation or controlling heat transfer. It is also still essential to build mockups for any glazing systems and test before installation and then down the road with commissioning to validate performance.

Translucent glazing materials, which are not transparent and which provide a diffused or even daylight in the interior space, can be used in a number of applications where a view is not necessary or even desirable. These materials can be glass, plastics, polycarbonates or acrylics, sometimes fiberglass, and are characterized by how they scatter visible light and prohibit the transmission of direct solar radiation through the glazing. As a result, glare and heat gain can be reduced by these materials with some additional advantages over transparent glazing materials. They also do not cause outward reflection into the environment like reflective glass which sometimes causes problems for adjacent buildings. Translucent plastics and fiberglass have lower thermal conductance, weigh less and cost less than glass, have better sound attenuation, but have a shorter life span, which makes them a good choice in particular situations, especially large glazed walls, skylights in top lighting and any areas where direct solar radiation could damage interior spaces or objects in them and where weight and initial cost are a factor. It is certainly possible to retrofit existing glazing systems as well by overlaying them with a translucent panel system where high heat gain and glare through single transparent glazing is a problem. See Figure 7-42.

Translucent glazing panels in walls and in skylights have been used very effectively for a long time. A few examples are shown in Figure 7-43.

RIGHT:
Figure 7-42.[90]

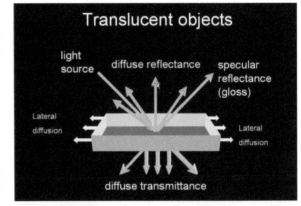

BELOW:
Figure 7-43.[91]

In addition to controlling glare, translucent glazing systems can be highly energy-efficient as well. For example, a system made by CPI, the Quadwall system (Figure 7-44), can deliver over 50% VLT with a SHGC of approximately .40.[92]

The systems shown in Figure 7-45 can be very cost-effective in a retrofit situation as well.[93]

Then how do all of these high-performance glazing technologies help the owner of an existing building? It is not all that often that existing buildings replace all of their windows or glazing, unless they are traditional buildings with punched openings. And it is very often cost-prohibitive. So, existing building owners often turn to other solutions such as shading systems and films to retrofit their existing glazing systems.

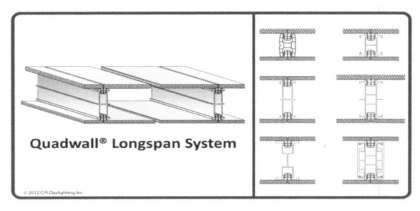

Figure 7-44

Figure 7-45

Shading Systems

Although shading systems can be exterior or interior, the high-performance and most efficient way to shade glazing from glare and solar heat gain is from the exterior. The reason for this is very straightforward: *exterior shading intercepts solar radiation before it contacts the surface of the glazing.* Once solar radiation has contacted the surface of the glazing, it is reflected, absorbed and transmitted and the energy transferred into the building must be acted upon by mechanical equipment and/or interior shades. This may seem intuitive but it is remarkable how often this fact is ignored in building design and retrofit, usually for reasons of aesthetics and cost. In the last ten years, however, fixed and operable exterior shading devices have begun to regain acceptance[94] in the design community as not only appropriate but also aesthetically pleasing on commercial buildings. Although their energy performance is proven and well-known, exterior shading systems do have their drawbacks—maintenance and upfront cost. These factors must be weighed carefully against long-term energy savings to determine the effectiveness of the strategy. Figure 7-46[95] illustrates the exterior shading principle well.

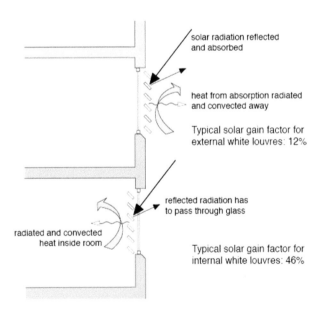

solar radiation reflected
and absorbed

heat from absorption radiated
and convected away

Typical solar gain factor for
external white louvres: 12%

reflected radiation has
to pass through glass

radiated and convected
heat inside room

Typical solar gain factor for
internal white louvres: 46%

Figure 7-46

A very useful short guide to daylighting strategy with shading is published by LBNL and shows numerous basic strategies.[96] The ASHRAE/IES Standard 90.1 "Energy Efficient Design of New Buildings Except Low-Rise Residential Buildings" has a major emphasis on shading in buildings, and this is to be integrated concurrently with assessment of high-performance glass.

An important study on the effectiveness of exterior shading (overhangs and fins) is "External Shading Devices in Commercial Buildings, The Impact on Energy Use, Peak Demand and Glare Control" by John Carmody and Kerry Haglund, sponsored by The Center for Sustainable Building Research at the University of Minnesota.[97] The building sites were computer modeled and included the cities of Minneapolis, Chicago, Houston, Phoenix, Los Angeles and Washington, D.C. Some of the conclusions from the report are that using exterior shading in conjunction with high-performance Low-E glass can reduce the overall energy use by about 11%, but reduces it almost 26% with clear insulated glass. At peak demand the reduction is about 20% with Lo-E and almost 46% with clear insulated. Reductions in glare are comparable, with approximately 35% with clear insulated[98] and approximately 40% with the Lo-E.[99] One of the interesting conclusions is that "it is possible to obtain equivalent performance to a low solar heat gain glazing with no shading device by using external shading devices with glazings that have higher solar heat gain characteristics."[100] This is good news for owners of existing buildings who have low-performing glass when replacement is impractical or cost-prohibitive.

Operable shading systems are generally superior due to their ability to adapt to changing conditions but are, of course, considerably more expensive. Exterior shading systems are categorized in Table 7-1.

Table 7-1

Fixed	Operable
Overhangs	
Fins(vertical louvers)	Fins
Louvers	Louvers (operable/retractable)
Screens	Screens (roller blind w/screen fabrics)
Awnings	Awnings

EXTERIOR SHADING SYSTEMS[101]

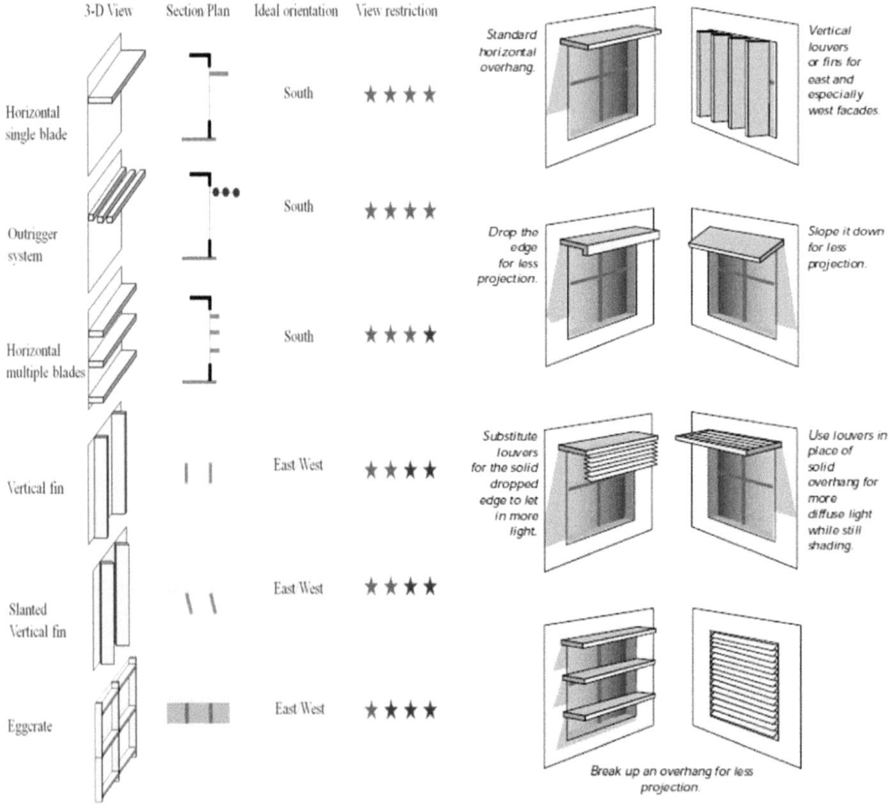

Figure 7-47. Exterior shading systems

The optimal combination (usually the most expensive) is a clear, insulated glazing system (preferably Lo-E) which allows a maximum of visible light and is shaded on the exterior by a system which allows good visibility, is operable and can respond automatically to changes in direct solar radiation, wind and overcast sky conditions. This can be achieved with operable louvers, blinds or retractable screens. There are dozens of manufacturers of these systems and we show some combinations of these products in Figure 7-48.

In new buildings, exterior shading systems are now usually integrated with daylighting controls which change electric light levels according to fluctuations in daylight intensity. Operable exterior systems

Tilak Competence Centre, Innsbruck, Austria. Colt Operable Vertical Louvers[102]

Zurich Airport. Colt Shado-Metal Horizontal Louvers.

Peninsula Residence. Automated Exterior Roller Shades w/solar screen fabric.[103]

C/S Solarmotion Retractable Exterior Venetian Blinds[104]

Figure 7-48

are automated and coordinated with the same. These systems are fully programmable, respond to external conditions such as wind as well as sun and can be integrated with BEMS. Figure 7-49 gives a typical layout.[105]

It is worth pointing out that since some exterior shading systems are lightweight and can be attached to existing structure, they can be retrofitted onto existing buildings. This makes exterior shading systems an excellent choice in many cases to upgrade a low-performing glazing system to high-performance.

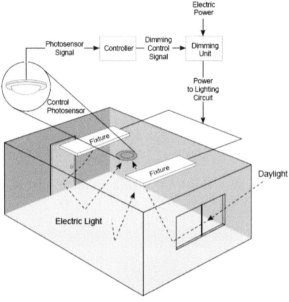

Figure 7-49

FILMS

Films for glass have been manufactured for a long time but have only in the past ten years become effective enough with spectrally-selective properties to qualify as contributing to a high-performance approach in a retrofit situation. There are a lot of retrofit situations where exterior shading systems are cost-prohibitive or not feasible for aesthetic or structural reasons. When combined with a high-quality interior roller shade system with solar screen fabric which cuts glare and some heat gain, some films can help improve a low-performing glazing system. Installing films on the exterior is more effective than on the interior for reasons we have already discussed, but is sometimes cost-prohibitive on multi-story buildings and carries with it other maintenance issues. However, films should be considered when other options are impractical. 3M manufactures a number of high-quality solar control films, including, for example, the Ultra Prestige 70.[106] According to 3M, this unique product can admit over 60% of Visible light, while cutting glare approximately 20%, is non-metallic, has a shading coefficient of .54, cuts UV by 99% and can make tempered glass behave like laminated glass in impact situations. With on-site validation, this product could be highly beneficial to building owners with single-pane glazing systems.

CASE STUDY

The New York Times building designed by Renzo Piano and built in 2006-7 provides an interesting case study for daylighting and shading design and implementation. The Times Company partnered with LBNL in 2003-4 to study a variety of strategies for increasing daylighting and reducing annual costs from electric lighting (see Figure 7-50).[107]

The RADIANCE lighting simulation program[108] developed at LBNL was used to calculate light and shade patterns that would fall on a proposed building given its orientation. (See Figure 7-51.)

Mockups were built and tested, including exterior tubular shade screens, interior automatic roller shades, and electronic dimming controls with T5 and T8 fluorescents. See Figure 7-52.

Daylight "harvesting" strategies are implemented using daylighting control systems that dim the electric lighting in response to interior daylight levels. For commercial applications, the light output of fluorescent lamps (T5 or T8) are varied by using electronic dimming ballasts. Photosensors, typically mounted in the ceiling, are used to measure the quantity of daylight in the space then determine the amount of dimming required to maintain the design work plane illuminance level. If

Figure 7-50

Figure 7-51

Figure 7-52
View of exterior façade of daylighting mockup.
Photo credit: Vorapat Inkarojrit, LBNL

daylight levels are more than adequate, the electric lights can be shut off. *Simulation studies indicate that annual energy use and peak demand can be reduced by 20-30% compared to a non-daylit building.*[109]

The selected technologies were commissioned at the mockup stage and the building was commissioned 5 years later to validate performance and savings.

The automated shading enabled lighting and cooling energy use reductions, and reductions in peak electric demand. Energy savings due to the shading system alone could not be determined in isolation but the reduction in annual electricity use due to the combination of all three systems was estimated to be 24% (2.58 kWh/ft²-yr) across a typical tower floor compared to a code-compliant building. *Annual heating energy use was reduced 51%. Peak electric demand was reduced by 25%.* The Times Company's investment in advanced energy-efficiency technologies was estimated to yield a 12% rate of return on their initial investment.[110] Figure 7-53 charts the estimated energy use.

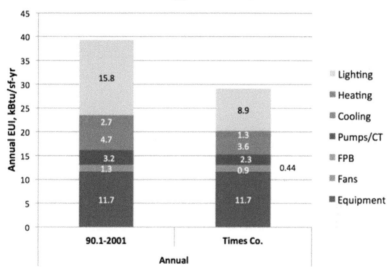

Figure 7-53
Annual end use energy comparison; baseline overhead versus calibrated Times Building model. Copyright: LBNL.

SOUND

Acoustical Criteria for Green Buildings
Acoustical performance is an important part of successful high performance building design. Excessive noise that interferes with an occupant's enjoyment of a building needs to be attenuated by careful

attention to various components incorporated into the structure. This is important because it is well known that excessive noise in the work-place can adversely affect performance and negatively impact produc-tivity.

Taking a systems approach, acoustical criteria must be applied to complete rooms, enclosures and structures and not just to the isolated elements comprising the building. By focusing on the performance expected from the system as a whole—foundations, walls, floors and from various inter-connections between these elements, it is possible to design a building that will function well as an overall system. Careful engineering design of the connections between the components is vital to extracting maximum acoustical benefit from all the materials used.

For significant noise reduction between two rooms, the wall (or floor) separating them must transmit only a small fraction of the sound energy that strikes it. The ratio of the sound energy striking the wall to the transmitted sound energy, expressed in decibels (dB), is called the transmission loss (TL). The less sound energy transmitted, the higher the transmission loss. In other words, the greater the TL, the better the wall is at reducing noise.

Acoustical Properties of Rooms

The three basic acoustical properties[111] of enclosures will largely determine the comfort level experienced by the occupants (see Figure 7-54).

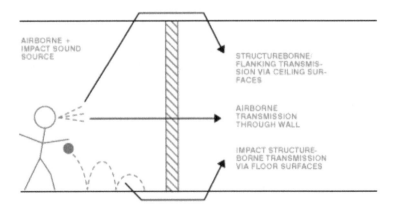

Figure 7-54

- Attenuation of the sound propagating between the enclosure and adjoining spaces;
- Background noise due to air-handling, plumbing and mechanical systems; and
- Reverberation of sounds within the room.

Sound Attenuation

All building components are ranked according to their ability to dampen airborne noise using the sound transmission class (STC). The higher the STC rating, the greater the level of sound that the material can attenuate. Different STC ratings can be specified as follows:

STC 25   Normal speech can be understood quite clearly.
STC 30   Loud speech can be understood fairly well.
STC 35   Loud speech audible but not intelligible.
STC 42   Loud speech audible as a murmur.
STC 45   Must strain to hear loud speech.
STC 48   Some loud speech barely audible.
STC 50   Loud speech not audible.

Building codes usually specify the minimum STC only for building components and not the acoustic isolation for complete pairs of enclosures. The objective is to design a "system" that provides the maximum acoustic attenuation between rooms.

To ensure that a desired sound attenuation between enclosures is achieved, designers commonly specify the apparent sound transmission class (ASTC). The ASTC represents the attenuation due to the combination of all acoustic paths between the enclosures in the completed building, not just the dampening due to the primary partition. Meeting a specification for ASTC requires a careful selection of components, using intelligently designed linkages between floors and walls and paying careful attention to small details during construction.

To reduce reverberation, sound-absorbing materials such as carpets, ceiling tiles and soft furnishings are commonly placed in an enclosure or on its surfaces. Calculating reverberation time for an ordinary room from the sound absorptions of the materials placed in it is a comparatively simple procedure.[113] Special enclosed areas, such as open-plan offices, lecture halls, concert areas and theatres need expert analysis of how sound propagates in the space and to specify the correct positioning of sound-reflecting and sound-absorbing materials.

The rating used to rank sound absorbing materials is the sound absorption average (SAA). The same material with different air cavities behind it can provide quite different SAA values, so a method of installation for testing must be specified that is similar to the intended final installation.

Sound Transmission

The standard approach to designing for the control of sound in North American multi-family wood-frame buildings commonly fails to consider the most important aspects of sound transmission. Building codes have usually only considered the sound transmission class (STC) rating of the assembly separating adjacent rooms within a structure or separating adjacent structures.

If the goal is to ensure a high level of sound insulation between units, then this approach can be considered to be inadequate because it does not consider flanking transmission which is structure-borne sound that bypasses the separating wall and travels through other building elements such as the floor, ceiling, or walls in contact with the separating wall.

Flanking transmission can be limited by the proper selection and design of the building as a complete system, meaning all the structural elements, including the foundations, paint, varnishes, finishes, junctions of partition walls, floors and ceilings. Reducing flanking by retrofitting is usually a much more expensive situation than dealing with the problem in the original design. In Figure 7-55, transmission paths between upper and lower rooms in a structure include both direct transmission through the separating floor and flanking transmission involving the floor and wall assemblies.

A sound generated within a room in a structure causes all surfaces of the surrounding space to vibrate. The vibrational energy is transmitted through the floor, ceiling, and adjacent walls and then through the floor/wall and wall/wall junctions to the corresponding surfaces of the adjacent room, where it is radiated as flanking sound. The control of vibrational acoustic energy requires not only a choice of properly engineered isolating assemblies, but also on the attenuation of the sound energy transmitted by all paths. The common term used to describe the transmission of sound via all paths is the apparent sound transmission class (ASTC).

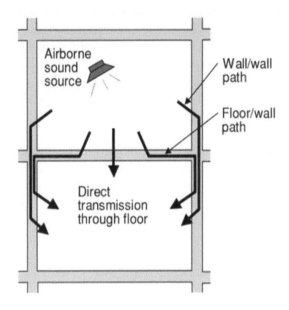

Figure 7-55

Sound transmission is often dominated by flanking through the surface of the floor, especially when this consists of a single layer of plywood. There are many possible approaches to reducing flanking through floor paths. The acoustical benefit of specific changes must be weighed and balanced against their cost to optimize the cost/benefit for the complete system.

Reducing the Transmission of Sound through Gypsum Board Walls

A major factor to consider in constructing walls to control sound transmission is the isolation of the gypsum board layers on each face of the wall. If at least one of the layers is not resiliently supported, or if the two faces of the wall are not isolated from each other, sound-absorbing material in the cavity is rendered ineffective. When the layers are isolated, sound transmission through the wall can be reduced by increasing the mass, the cavity depth and the amount of sound-absorbing material. To provide the needed isolation, walls can be constructed of double studs (wood or steel), staggered studs, non-load-bearing steel studs, or wood or load bearing steel studs with resilient channels.[114]

The type of sound-absorbing material has a relatively minor effect on the ability of the wall to control sound. There are clear benefits associated with increasing the spacing between studs and resilient channels. It is found that the farther apart they are, the better the sound reduction. Usually the STC increases by one or two points when going from 400 mm to 600 mm spacing. Figure 7-56 shows some examples of strategies with a product called Enermax.[115]

Reducing the Transmission of Airborne Sound through Floors

The parameters that can be varied include joist type, depth, material stiffness, spacing, type and thickness of sound-absorbing material, type and arrangement of furring used to support the gypsum board, type and thickness of the subfloor and type and thickness of the gypsum board. It is found that the key factor in increasing sound isolation in

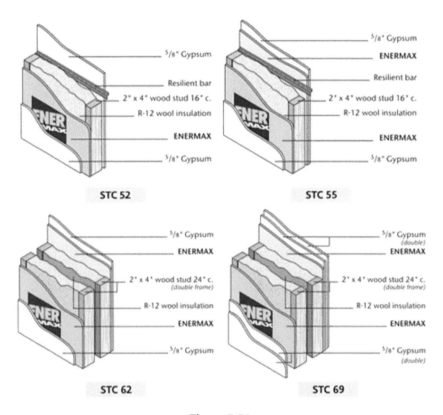

Figure 7-56

joist floors is the independent or resilient support of the gypsum board ceiling from the joists. If the gypsum board is not supported in this way, sound-absorbing material in the floor cavity is rendered ineffective. The most common way of resiliently supporting gypsum board ceilings is to use 25 ga. resilient metal channels.

In floors without resilient metal channels, acoustical energy transfers from one surface of the floor to the other primarily through the joists. Thus, adding sound-absorbing material to the cavity has little effect on the sound isolation.

While furring systems provide an improvement over the direct application of gypsum board, they are still too rigid and not as effective as resilient metal channels, where the flange supporting the gypsum board is free to bend. Adding a thickness of concrete topping, for example, to a basic floor assembly increases the sound attenuation because of the increased mass.

In joist floors with resilient metal channels attached to the underside to support the gypsum board ceiling the most important factor is the total mass per unit area of the subfloor and ceiling layers. The mass of the joists is not a significant factor. Important factors include the thickness of the sound-absorbing material, the arrangement of the resilient metal channels and the depth and spacing of the joists. Putting sound absorbing material in the cavity of a joist floor with a ceiling that is directly attached to the joists provides no significant improvement in sound attenuation.

In summary, the influence of sound-absorbing material in a floor cavity can be summarized as follows:

- Increasing the thickness of the sound absorbing material increased the STC while decreasing the thickness decreases it.

- Each change in thickness of about 65 mm changes the STC by 1 point.

- For the same thickness of material, rock and cellulose fibers give an STC that is higher than that of glass fiber by about 1 point.

- To maintain the fire resistance of ceilings consisting of single layers of gypsum board, it is possible to add pieces of resilient metal channel to support the butt ends of the gypsum board, reducing the STC by 1 – 2 points.

- Increasing the depth of the joists by about 100 mm increases the STC by 1 point.

- Increasing the spacing between joists by about 200 mm increases the STC by 1 point

- For practical purposes, the type of joist (solid wood joists, wood I-joists, wood trusses, and steel joists) does not significantly affect the sound isolation.

- Attaching the subfloor to the joists using both construction adhesive and nails gave the same sound isolation as attaching it using only screws.

- Changing the gauge of steel joists through a range of 14 to 18 has little effect on the STC.

- Changing the width of the I-beam or truss flange in contact with the subfloor has little effect on the STC.

- Placing resilient metal channels between two layers of gypsum board in the ceiling significantly reduces the sound isolation.

Figure 7-57 is an example from Ridge Development Corporation.[116]

Sound Attenuation in Concrete Slab Floors

The sound attenuation provided by concrete floor slabs is influenced mostly by the mass of the slab. A gypsum board ceiling suspended resiliently below a concrete slab can significantly increase the sound damping. The magnitude of this increase depends on the mass of the gypsum board, the depth of the cavity between the board and the concrete slab, and the thickness of the sound-absorbing material in the cavity. A cavity depth that is made too small can reduce the sound isolation and the STC. Increasing the cavity depth and placing sound-absorbing material in it increases the sound isolation.

Reducing the Transmission of Impact Sound through Floors

Impact sounds, such as those created by footsteps, or by the dropping or moving of objects, can be a source of great stress to the occupants of buildings. The character and level of impact noise generated in the living space below depends on the object striking the floor, on the structure

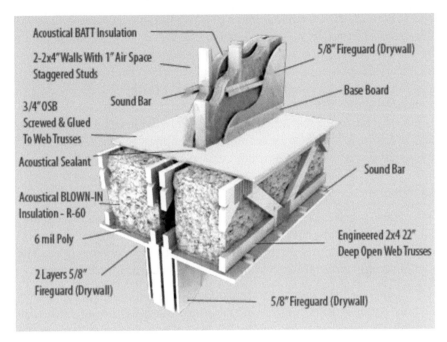

Acoustical BATT Insulation

2-2x4"Walls With 1" Air Space
Staggered Studs

5/8" Fireguard (Drywall)

3/4" OSB   Sound Bar
Screwed & Glued
To Web Trusses

Base Board

Acoustical Sealant

Acoustical BLOWN-IN
Insulation - R-60

Sound Bar

6 mil Poly

Engineered 2x4 22"
Deep Open Web Trusses

2 Layers 5/8"
Fireguard (Drywall)

5/8" Fireguard (Drywall)

Figure 7-57

of the floor assembly and on the floor covering. For each floor assembly, it is possible to measure the impact noise level and generate a relative rating known as the impact insulation class (IIC). The higher the IIC, the better the attenuation of impact sound, with 50 usually considered the minimum rating for occupant satisfaction in residential buildings.

Impact sounds on concrete slabs finished with a hard surface such as ceramic tile can be described by rather unscientific terms such as "clicking," "clacking" or "taping." Most of the energy of such sounds is emitted at high frequencies. For typical concrete floors, the IIC is relatively low and is determined by the high frequency acoustic energy. Concrete floors finished with tile or other hard materials are responsible for many of the complaints about noise produced by footsteps and the moving of furniture.

Impact sounds on lightweight joist floors, on the other hand, are usually described by equally unscientific terms such as "thumping," "booming" or "thudding." Most of the energy of these sounds occurs in the low frequency range. With typical joist floors, more low-frequency sound is transmitted than in the case of concrete floors.

There are two principal ways of providing an acceptable finished floor surface, namely, by use of flexible layers such as vinyl or carpet or by use of "floating floors," consisting of a slab of rigid material supported on a resilient mat or pads, or by use of various combinations of these. The kind of topping combined with the type of floor structure, will have a significant effect on the reduction of impact sound. Concrete slabs finished with hard surfaces such as ceramic tile, marble or hardwood have low IIC ratings. Satisfactory cushioning of impacts can be accomplished by installation of a resilient upper surface or a floating floor.

In Summary

- Hard-finish flooring materials such as ceramic tiles that are bonded directly to concrete slabs will not significantly improve upon the impact sound attenuation achieved by the concrete itself. The flooring material selected must be one that cushions the impact.

- Concrete slabs finished with wood flooring produce an only slightly better impact performance than uncovered concrete. Although the IIC may be improved by the use of adhesive used to bond the wood to the concrete, without the use of a resilient layer under the wood, the impact attenuation of the floor will be unsatisfactory.

- Placing wood flooring on top of a resilient layer will provide an acceptable attenuation of impact noises. The IIC value obtained depends mostly on the resilient material used. Effective materials that are used include shredded or foamed rubber, foamed plastic or cork mats. Increasing the thickness of the resilient material will generally increase the IIC.

- A floating floor can consist of a top layer of wood supported on strapping with a layer of fibrous material.

- A concrete top layer over a layer of glass or mineral fiber gives an even better IIC rating than a layer of wood on strapping. The thickness of the concrete layer may range from 30 mm up to 100 mm. Resilient pads consisting of cork, rubber or shredded recycled tires and fibrous batts made of materials such as glass fiber can be used in place of fiber boards

- A very high IIC rating can be achieved by use of the cushioning effect provided by carpet with underlay. Increasing the thickness of the concrete slab to 200 mm would increase all the IIC ratings by three to four points.

- Gypsum board ceilings suspended independently from concrete slabs offer greater resilience and provide an excellent additional method for increasing impact sound attenuation.

- Increasing the mass of the gypsum board, the depth of the cavity, or the amount of sound-absorbing material all increase the IIC relative to that of a bare concrete slab. The increase can range from four or five points to more than 30.

Effect of Different Types of Electrical Outlet Box

Poorly placed electrical outlet boxes can significantly decrease the sound isolation of gypsum board walls—a decrease of up to 6 STC points. This can be minimized by:

- Making sure that (untreated) metal boxes are offset by 400 mm or more in adjacent stud cavities rather than being placed within the same stud cavity,

- Using plastic vapor-barrier boxes,

- Employing retrofit techniques that emulate the attributes of plastic vapor-barrier boxes. The presence of sound-absorbing material in the cavity helps to further minimize the effect of poor placement, especially when installed so that the material blocks the line of sight between the boxes.

Acoustical Design for Open-Plan Offices

The open-plan office has become the predominant type of office space for a wide range of work-related activities. In the majority of offices, older designs with stand-alone panels and furniture have typically been replaced by modular workstations and cubicles. This type of office is considered to be cheaper to construct and reconfigure than other types. However, there are factors, such as the lack of privacy and increased noise, that have negative effects on office workers and that need to be mitigated by appropriate design to obtain an acceptable level of noise isolation.

Noise isolation is often referred to as speech privacy because intruding speech sounds are often adversely affect many individuals ability to concentrate on their work. Speech privacy is related to the level of unwanted speech sounds from adjacent workstations relative to the level of more constant ambient noise. Reducing intruding speech sounds or increasing background noise levels can both improve speech privacy, although at some point the noise level may itself become a problem.

Some definitions that are required include:

- **Speech Intelligibility Index (SII)** is a measure derived from the signal-to-noise level differences in each frequency band, where the differences are weighted according to their relative importance to the intelligibility of speech. These weighted signal-to-noise level differences are summed to obtain a SII value that is between 0 and 1. This measure indicates the expected speech intelligibility in particular conditions: An SII of 1 indicates conditions in which near perfect speech intelligibility is expected, whereas an SII close to 0 indicates conditions in which near perfect speech privacy is expected. SII has replaced the articulation index (AI) and has values that are approximately 0.05 larger than corresponding AI values.

- **Sound Absorption Average (SAA)** is an average of the absorption coefficients in the 1/3-octave frequency bands from 250 Hz to 2.5 kHz. It replaces the older noise reduction coefficient (NRC) measure and has similar values for the same material.

- **A-weighted sound level (dBA)** is a simple measure that weights and sums the contributions of sounds at different frequencies to approximate the total loudness experienced by listeners.

- **Sound Transmission Class (STC)** is a single number rating of the sound transmission characteristics of panels. Higher numbers indicate greater attenuation of the transmitted sound.

There is a reciprocal relationship between speech privacy and speech intelligibility—the lower the speech intelligibility the greater the speech privacy. To achieve adequate speech privacy from intruding speech sounds, speech intelligibility scores must be low. Speech privacy and speech intelligibility are both related to the loudness of the speech compared to that of the ambient noise and both use measures of the signal-to-noise level difference, where intruding speech is the "signal" and the general ambient noise is the "noise."

Sound Paths Between Workstations

As mentioned in the previous section, SII is widely used as a measure of speech privacy. An SII ≤ 0.20 is considered to provide normal or acceptable speech privacy in open-plan office situations[117].

Since speech privacy is related to the speech-to-noise level difference, reducing speech levels in the open-plan office can improve speech privacy. Office etiquette that encourages occupants to keep their voices down is an essential starting point. Re-locating prolonged or animated discussions to closed meeting rooms is also desirable. It is difficult to obtain acceptable speech privacy if the general ambient noise level is too low. On the other hand, if the ambient noise level is too high, it can be annoying and cause people to talk more loudly. There is a narrow range of ambient noise levels that can mask intruding speech sounds from adjacent workstations without being disturbing. For this reason, successful open-plan office designs typically use electronic masking-sound systems.

Electronic masking-sound systems can be designed to provide close-to-ideal noise levels to mask speech sounds and enhance privacy, without being disturbing. The masking noise should be adjusted to sound like ventilation-system noise and it should be evenly distributed throughout the office (variations should be less than 3 dBA). A masking-noise level of 45 dBA is judged to be optimal, with a level of 48 dBA considered the maximum acceptable level (14). These levels roughly correspond to ventilation noise ratings, commonly used by heating and ventilation engineers.

Masking-sound systems include those with centrally located electronics and those with distributed units. Manufacturers of both types claim various practical advantages. Although the sound propagated into the ceiling void (the space between the suspended ceiling and the structural floor) may aid the even distribution of the masking sound in the office itself, it gets modified when transmitted through the ceiling tiles and lighting fixtures. This can lead to localized areas of higher sound levels in the office space.

More recently, masking-sound systems with loudspeakers mounted in the ceiling tiles and on workstation panels have been proposed in order to provide better control over the masking sound. The installation of masking-sound systems is best left to experienced professionals.

Reducing Speech Propagation

The acoustical design of an open-plan office can be complex because of the many different sound paths that need to be considered. Sound can be reflected from the ceiling, diffracted (or bent) over the top of a separating panel, or transmitted through the panel. One can conveniently calculate the effect of these various sound paths using acoustical design software[118] that is based on a complex mathematical model[119] of sound propagation between adjacent workstations (see Figures 7-58 and 7-59).

Ceiling Absorption and Panel Height

Such calculations show that the most significant paths are those that reflect sound from the ceiling and diffract sound over the separating panel. It is therefore essential that the combination of ceiling absorption and separating panel height be adequate. Designers can refer to tables listing values of average ceiling absorption that are a function of separating panel height.

The size of the workstation is also important for achieving adequate speech privacy. If a workstation is reduced then the speech privacy will also be reduced. To achieve acceptable privacy in a workstation that is smaller, it is possible to increase the ceiling absorption and separating panel height to the maximum possible. It should be noted that workstations smaller than 2 x 2 m cannot be designed to provide acceptable speech privacy.

Many other design parameters influence the level of speech privacy in an office. These include:

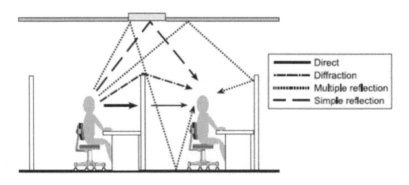

Figure 7-58

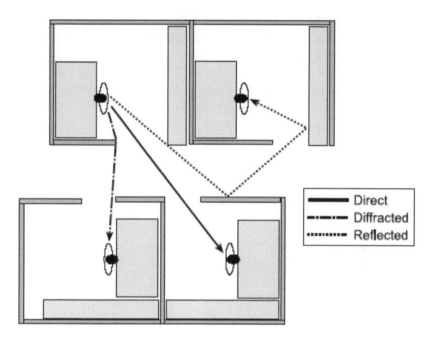

Figure 7-59
Examples of direct, diffracted and reflected sound paths between workstations

- **Panel Characteristics**: To control speech sound transmission through workstation panels, the panels must have an STC rating of at least 20.

- **Wall Treatment**: Where there are large areas of wall, they should be treated with sound-absorbent material

- **Floor Treatment**: Floor absorption and ceiling height generally have very small effects on SII. In most cases, varying these parameters changes SII by no more than 0.01. However, floors should be carpeted to minimize activity noises and sound propagation through gaps at the bottom of workstation panels. When the floor is carpeted, gaps of up to 25 mm have negligible effects on sound propagation between workstations.

- **Light Fixtures**: Located in the ceiling, these can degrade the speech privacy of the open-plan office. The magnitude of the effect depends on the type and location of lighting units. Lights with a flat plastic or glass surface produce strong unwanted acoustic reflections and are most troublesome when located directly above the separating panel between two workstations. The change in SII is greatest when these lights are installed in a highly absorptive ceiling. The use of parabolic louver (open-grill) lighting fixtures has a less negative impact on speech privacy, but still reduces the effectiveness of a highly absorptive ceiling.

- **Workstation Layout**: Sound paths in both horizontal and vertical planes should be examined to identify possible direct or reflected paths between workstations. Vertical surfaces outside individual workstations should be made sound absorbent to prevent strong sound reflections between workstations. Where there is a direct line of sight between two nearby occupants, the layout should be changed to eliminate the direct sound path.

- **Window/Panel Interface**: When workstations are located next to windows, it is often difficult to avoid having gaps between the panels and the window. Such gaps can allow for strong sound reflections between workstations, greatly reducing speech privacy. Efforts should be made to completely fill such gaps.

Team-Style Work Spaces

Team-style open office spaces usually resemble a large cubicle with multiple occupants. Achieving acceptable privacy between such spaces is the same problem as between individual cubicles. Although intended to provide easy communication between team members, it is at times desirable to have some speech privacy between occupants of the team-style space. It is important that the ceiling be as highly absorptive as possible. Team-style spaces are not suitable for individual work requiring concentration but are most suitable where the work involves considerable interaction throughout the day.

There are essentially no barriers between occupants of the work area, but there are barriers or panels separating them from adjacent work spaces. In this type of space, the occupants are in full view of each other and clear speech communication between occupants can occur if they desire it. Although it may not be possible to change the basic concept, some details of the design can help to improve acoustical conditions.

- All panels and other large surfaces should be sound absorptive.

- Small, low barriers can be used to break the line-of-sight between adjacent occupants and to improve speech privacy without detracting from the open feeling of the workspace

- Occupants should be oriented so that, when working independently, they are facing away from each other.

- The ceiling should be highly sound-absorbent

- The distance between occupants should be maximized

Attenuation of Sound by Green Roofs

Because of their large number of ecological and economic advantages, vegetated roof tops (green roofs) have become an increasing popular addition to green buildings. Green roofs have important acoustical benefits as well. Green roofs have a higher weight resulting in an increased sound insulation of the roof system. This could lead, depending on the geometry of the building, to strong reductions in indoor noise levels.

Green roofs can also be used to successfully reduce road traffic noise exposure, which is the main source of noise annoyance in urban areas. Diffracting waves may reach the backside of a building or a roof apartment (penthouse) after shearing over the roof. Despite the shielding caused by the building, noise levels may still be unacceptably high at these locations. The typical substrates used for green roofs are of high porosity and thus allow sound waves to enter the growing mediums. Attenuation of the sound occurs because of the large number of interactions between sound waves and substrate particles. As a result, less noise will reach the back façade or the roof apartment compared to the situation with a non-vegetated roof that is most often made of a highly reflective material such as wood or concrete.

The attenuation by a green roof increases strongly with increasing sound frequency. Lower frequencies are incapable of interacting significantly with the growing medium, and consequently little attenuation occurs. For frequencies of 1000 Hz or more, an increase in shielding of up to 10 dB is observed by the presence of an extensive flat green roof covering 80% or more of a roof surface, compared to exactly the same building geometry with a fully rigid flat roof. This is important because typical traffic noise consists of sounds at frequencies of 1000 Hz or greater.

It is found that substrate layer thickness is an important design parameter. For noise at frequencies of 1000 Hz, the optimal thickness is about 0.10 m. With decreasing noise frequency, this optimum will gradually shift towards thicker substrates. Economics will limit the maximum thickness of substrate that is practical. The use of shielded building façades to reduce road traffic noise is therefore an important supplement to the noise reduction provided by extensive green roofs.

The fraction of the roof covered by growth medium will determines the improvements that can be obtained. The predicted attenuation, relative to a rigid roof, increases linearly with the green roof fraction. This means that a tilted green roof leads to a larger (relative) improvement in shielding than a flat green roof, since there is more interaction between sound waves and the substrate.

It can be concluded that both extensive and intensive green roofs have the potential to significantly reduce sound intensity levels, and that replacing a rigid roof by an optimized green roof, such as a saddle-backed roof for example, improves shielding from road traffic noise up to 10 dBA. For the average occupant, such a reduction in sound intensity level corresponds to halving of the loudness.

Sound Attenuation by Energy Efficient Windows

First and foremost, high-performance windows are designed to be energy efficient. They can offer significant savings on HVAC and lighting costs. In addition, the heavy gas used in high-performance windows (such as argon), which is designed to reduce conductive and convective heat losses, also has the secondary effect of reducing sound transmission from the exterior to the interior. The greater the number of panes, the better the sound absorption. Sound attenuation is complex, depending on frequencies and other factors but significant acoustical benefits will be realized with the use of high-performance windows.

When considering windows, there are generally three options available for maximum possible sound attenuation.

- Use of laminated glass,

- Employing a wider airspace between the panes,

- Using two panes of different thickness in the Insulating Glass Unit (IGU), or

- Using a combination of all three of these approaches.

As an extreme example, windows used in airports usually employ laminated glass on both sides of the IGU in an aluminum frame and with a "maximum" airspace between the panes. Obviously, with this application, the primary concern is sound attenuation and energy efficiency is only considered a secondary benefit. This is important because the width of the airspace chosen in a particular window design and the choice of framing material will impact both the acoustic and energy efficiency of the unit.

Many energy efficient window units use triple pane glass. If considered for use in sound damping, triple pane units provide a slight improvement over standard double pane units at lower frequencies due to the additional density provided by the additional layer of glass. However, overall there is no difference in STC rating between triple and double pane provided that the overall airspace between the panes is constant between the two constructions. As a numerical example, consider a triple pane window with two 1/4" airspaces and a dual pane unit with a single 1/2" airspace, both using 1/8" glass, then the STC will be identical if the IGUs are the same dimensions.

Using one thicker (3/16") and one thinner (1/16") pane of glass in an IG construction will help deaden sound because each pane is "transparent" to a different frequency and each will then attenuate the frequency that the other pane allowed to pass.

A fixed, non-operable, window will often show significant improvement over an operating window. Some acoustic performance numbers are provided below.

- A 1/4" monolithic pane has an STC of 31

- A 1/4" laminated pane has an STC of 35

- A 1/2" (overall width-airspace width is 1/4") IGU has an STC of 28, lower than a single sheet of 1/4" glass

- Increase the airspace in the IGU to 3/8" from 1/4" and the STC increases to 31, or the same as a monolithic 1/4" pane

- In the same laminated glass/IGU make up, if the airspace is increased to 1/2," then the STC increases to 39

- With an increase of the airspace to 4" and using laminated glass on both sides of the airspace the result is an STC of 53 (Airport-Standard)[120]

SOUND—Conclusion

In an attempt to make green buildings as energy efficient and high-performance as possible, designers almost universally also succeed in making them acoustically efficient as well. It has been shown that many energy conserving features incorporated into walls, roofs, windows and floors also lead to the creation of a much quieter, and consequently a more pleasant and less stressful internal space.

ENVELOPE—PRINCIPLE, FORM & FUNCTION

Until fairly recently (the last 150 years), envelopes have usually been constructed with local materials and have been designed to work with the existing local climatic conditions.

After looking at principles of energy transfers and the function of building components, we can identify, define and summarize some fundamental types of envelopes which have used or could use these principles and functions. It shows that the evolution of high-performance buildings borrows from the past while leveraging the materials science of today. These categories are not hard and fast and allow for transferability between them. There are over a hundred case studies of high-performance and zero-energy buildings available from WBDG[121] and DOE[122]. We'll consider a few examples which illustrate the principles listed in Table 7-2. Some of the examples are of buildings with which I have personal experience.

Heavy Mass

Heavy mass envelopes were used in traditional buildings, although the principle has been revived recently, mainly in residential and some small commercial buildings.[124] A good ancient example of (a) is the Pantheon in Rome, built circa 27 B.C. by Marcus Agrippa, and later rebuilt by the Emperor Hadrian circa 126 A.D. Section.[125] The walls of the Pantheon are in some places 13 feet thick. See Figure 7-60.

Why use an example from ancient architecture to discuss high-performance buildings of today? Because the principles of heavy mass envelopes are important from the standpoint of strategy when it comes to energy flows. Modern thin-wall construction using engineered materials benefits from these principles of Retard-Retain.

I visited the Pantheon when I was in Rome one summer in early

Table 7-2

[1] Heavy-Mass	**BARRIER** **PRINCIPLE: Retard-Retain** Thick envelope, usually masonry often with restricted or single-orientatation glazing/daylighting, passive energy management
[2] Light-Mass[123]	**BARRIER/MEMBRANE** **PRINCIPLE: Retard-Retain • Reject-Remove** Thin envelope, masonry veneer, frame or curtain walls with larger glazed areas, orientation can be passively solar designed for optimization but usually relies mainly on active methods for energy management such as HVAC.
[3] High-Performance	**BARRIER/MEMBRANE/COLLECTOR PRINCIPLE:** **Retard-Retain • Reject-Remove • Absorb-Translate** Thin or thick envelope which can act as a combination barrier/membrane/energy converter. Design can optimize orientation, thermal efficiency, daylighting, collection, and substantially reduces the requirements for mechanical and electrical equipment. Can have net-zero as a goal.

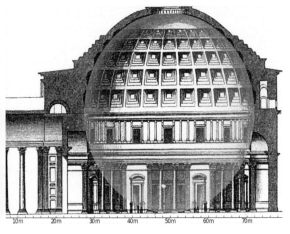

Figure 7-60[126]

July when it was over 90°F outside. Although I didn't measure it, when I got inside the building, the temperature was at least 15° cooler (quite comfortable) and the inside of the structure was bathed in daylight from the oculus at the top. The mass of the walls retarded the heat flow from outside to inside. Of course heavy-mass wall envelopes are nowadays impractical and cost-prohibitive to build except in particular circumstances, and this is why thin-wall envelopes have evolved as we have seen to mimic some of the retard-retain functions using *layers* of insulating materials, air spaces and reflective coatings.

Light Mass

The evolution of envelopes to light-mass occurred as a result of the development of materials such as milled wood, float glass, steel, aluminum and plastics, concurrent with the deployment of fossil fuel energy sources such as coal, oil and natural gas. A light-mass envelope also retards and retains energy flows, but it does this with the *resistance value (R-value)* of the wall components rather than the *thermal mass effect* of the heavy mass wall. A light mass envelope is engineered such that it can be erected much faster and more cost-effectively than a heavy-mass envelope, but it depends on the ability to *reject and remove* unwanted energy flows using mechanical heating, ventilating and air-conditioning which require remotely supplied energy. Thus the general strategy with energy flows in light-mass envelopes is to increase the R-value per unit thickness of the wall components, preferably without substantially thickening the wall and increasing cost.

Light mass envelopes, even if they have not maximized the R-values (or U-factors) of the components, can benefit greatly from proper orientation and passive solar design, which nevertheless reduces the mechanical and electrical lighting requirements. A very good example of high-performance design without high-performance building components (a candidate for glazing retrofit) is the Meadows Building built in 1954 and located in Dallas, TX (see Figure 7-61).

The rectangular Meadows Building orients the main facades with glazing to the north and the south, with east and west facades with small, narrow punched openings. This minimizes envelope exposure to direct solar radiation, especially important in that climatic zone. The south elevation shows deep overhangs/porches over the south facing glazing to also minimize direct solar radiation on the wall surfaces and the windows. The building combines numerous materials:

North Elevation

South Elevation

Figure 7-61.

Figure 7-62[127]

stone, metal, concrete and glass in a way that maximizes visibility and reduces energy transfers. This is classic passive solar design with a light mass envelope which also optimizes daylighting in the building. Ideally these principles would be used in the orientation of any building in a hot climate. I have visited a number of offices in the building on both the north and south sides of the building. Although the building was originally fitted with single glazing, and of course utilizes mechanical equipment, the interior spaces were comfortable and well lit with daylight.

But the vast majority of light-mass envelope commercial buildings have become glass curtain-wall structures in the last 60 years. Some of this has to do with cost, some of it with aesthetics, and almost none of it, quite frankly, with the fundamental principles of high-performance building. (See Figure 7-62.)

The energy performance of a glass curtain wall is almost entirely dependent on the glazing system and that is a little too much to ask of those materials. This type of envelope rarely leverages orientation, controlled daylighting and shading and is therefore focused on the reject-remove principle and is highly reliant on HVAC equipment, electric lighting and a low cost of electricity. This makes a lot of these buildings ideal candidates for high-performance envelope retrofits (well maybe not the Bauhaus!).

High-performance

As we have already seen in Chapter 3, ASHRAE Standard 189.1 and the IgCC give detailed prescriptions for achieving high-performance goals with regard to the envelope. But these are largely numerical targets based on ratings for R-values and solar heat gain coefficients (SHGC) of components. They do not include principles or strategies for employing design, orientation or specific products or materials.

Choosing which HP buildings to talk about is very difficult as there are hundreds of excellent examples. So I am choosing some examples of smaller HP buildings here which have [1] very low energy consumption per sq. ft., [2] have some ingenious approaches to high-performance features, [3] are closer to average construction costs per sq. ft., [4] and are not heavily cost subsidized. I am also selecting from different climate types: warm/cold/dry, temperate, hot/cold/humid, hot/humid, etc. The High-Performance Buildings Database[128] and the

Whole Buildings Design Guide[129] together list more than a hundred representative case studies of buildings which can be examined. There are a number of very impressive Class A buildings in these databases with extraordinary architecture and cutting edge high-performance features. These are very important from the standpoint of showing us what can be achieved with significant technologies and outstanding design. However, they are new construction, represent a very small percentage of the overall market of commercial, industrial and institutional buildings and tend to have very high construction costs per sq. ft. This means they fall outside of what is feasible for the vast majority of existing buildings and for new buildings which do not have Class A budgets. Since great detail is available about the following cases on-line, I will simply review the major points of the high-performance nature of the projects.

Warm/Cold/Dry—Taos, NM, Residential

The so-called EarthShips of northern New Mexico (and now getting built around the world) are an outstanding example of successful modern heavy-mass envelopes utilizing passive solar architecture and high-performance features. Of course their building techniques cannot be replicated in high-density urban areas or in multi-story buildings, nevertheless they do illustrate HP features which approach net-zero, zero waste, and include the use of local and re-used building materials. See Figures 7-63 and 7-64.

The walls are typically constructed of rammed earth using local and recycled materials such as radial tires and sometimes cans and bottles, creating a large thermal mass, which absorbs and conducts energy very slowly. These are hybrid buildings with thick and thin envelope sections, which employ all three principles of [1] retard-retain, [2] reject-remove and [3] absorb-translate. I happened to stay in an EarthShip, the Dobson House[132] (a bed & breakfast) built by a former engineering professor from the University of North Texas. It was March and the outside air temperature in the morning and in the evening could be quite cold at that time of year, but the inside of the building where we stayed was warm and comfortable with ample daylighting through sloped glass walls. In this sense, the EarthShip *retards and retains* with heavy mass, but it also *absorbs and translates* by collecting solar energy, storing it in the building and converting it to electricity as well. It minimizes the requirements for *reject and remove*.

Figure 7-63[130]

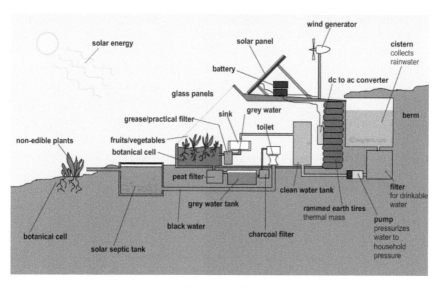

Figure 7-64[131]

Temperate—San Francisco, CA, Restaurant, Industrial, Commercial
355 11th St. Aidlin Darling Design

This is a 3-story, 14,000 sq. ft. building, mixed use, and an adaptation of a former turn-of-the-century industrial building in an historic district. The design and construction team faced a number of challenges, especially with regard to the envelope, which because of historic district restrictions could not be significantly altered in outward appearance. So an ingenious strategy of installing a perforated metal screen on the outside of the building which provided daylighting, shading, natural ventilation and allowed for additional fenestration behind it solved some of the problems. The original post and beam framing was maintained, the building has a green roof and also, subtracting for cooling and fan energy (unused, there is no cooling system), approximately 37% of the building's electricity is produced with photovoltaic panels. The building received a USGBC LEED-NC rating and received its energy credits in accordance with ASHRAE 90.1. The total EUI for the building is 19kBtu/sq./yr. The cost per sq. ft. is unavailable but the project was privately financed.[133] See Figures 7-65 and 7-66.

Hot/Cold/Humid—Chicago, IL, Commercial Office-Non-Profit
Center for Neighborhood Technology(CNT),
Farr Associates, J.T. Katrakis & Associates

The Center for Neighborhood Technology has been in existence since 1978 and is dedicated to sustainable development and community access. The CNT is housed in a 3-story, 15,000-sq.-ft. renovated 1920 building which is LEED Certified NC-Platinum and has an Energy Star rating of 89. The renovation was self-financed 90% with 10% in grant funding. The CNT team wanted to establish that LEED Platinum rated construction could be accomplished within a "cost comparable to conventional rehab." So the team used off-the shelf technologies, super-insulated the envelope and used high-performance windows, achieving a construction cost of $82 sq./ft. "The renovation reused 100% of the existing building's structural material and 90% of its shell components. Over 70% of the building's materials are recycled."[136] The CNT developed its own building performance tool and the commissioning team has monitored the energy consumption of the building since completion in 2005. Energy use is approximately 44% less than comparable buildings benchmarked for the same purpose, with an annual average EUI of about 53kBtu/sq.ft. A photovoltaic array provides about 5% of the electricity required. The CNT has an interesting ice-storage tank used with the cooling system which contains water-filled plastic balls suspended in glycol. When the balls freeze, the cooled glycol runs through pipes in the building to help cool it.[137] See Figure 7-67a & b.

Figure 7-65[134]

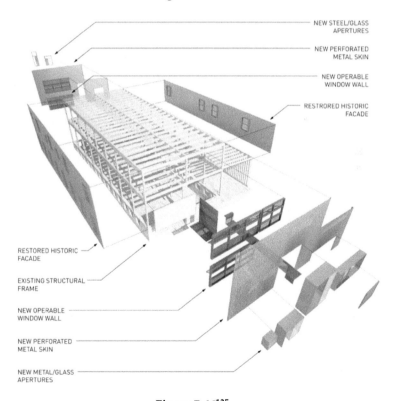

Figure 7-66[135]

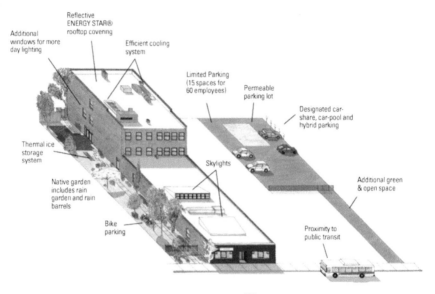

Figure 7-67a[138]

Figure 7-67b[138]

Hot/Humid—Gainesville, FL, Higher Education
Rinker Hall at the University of Florida, Croxton Collaborative Architects

Rinker Hall is a 3-story, 47,300 sq. ft., leadership facility within the University of Florida's College of Design and Construction. It holds a LEED NC-Gold rating and is oriented on a pure north-south axis which maximizes low-angle daylighting for this latitude and climate. See Figure 7-68.

Figure 7-68[139]

"An interactive and open design process characterized the design of Rinker Hall from its start. Two three-day design workshops on-site with faculty and students from the School of Architecture and School of Building Construction, university personnel, the construction manager, and the design team forged open communication from day one."[140] The project's construction-management team was made up mainly of graduates of the School of Building Construction. It is obvious that the approach was integrative, collaborative and inclusive of future occupants, a very innovative approach. High-performance strategies accounted for $182,000 of the total $6.5M construction, approximately 3%, and the building was built for $137.50/sq. ft., a very competitive number for comparable buildings. "Energy conserving measures include shade walls on the west and south facades, a high-performance envelope, lighting controls, a high-albedo roof, and energy-recovery ventilation. The high-performance envelope includes a metal panel vented rainscreen system, an air-infiltration barrier, a thermally broken aluminum storefront, and high-performance glazing. The exterior wall system was carefully designed to balance moisture, thermal loading, and daylight."[141] See Figure 7-69.

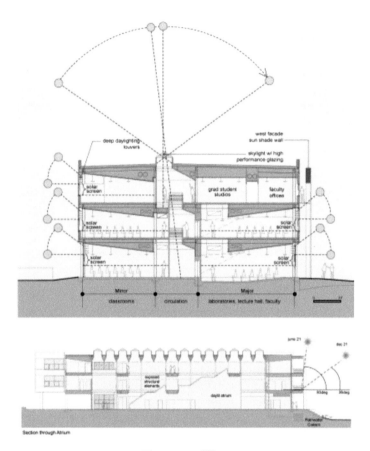

Figure 7-69[142]

Rinker Hall was expected to use 57% less energy than a comparable, baseline building designed in minimal compliance with ASHRAE 90.1-1999. Modeling with DOE-2.1e predicted that an enthalpy wheel (energy recovery ventilation—ERV)[143] would cut peak load demand by 22%. See Figure 7-70.

"Of the 55% peak load reduction, 40% of the reduction is achieved by the project's 8-foot-diameter, 12-inch-thick enthalpy wheel (ERV), and 60% is achieved through optimization strategies such as daylighting, the efficient envelope and roof, occupancy sensors, shading, and the west thermal wall."[144]

Rinker Hall is an excellent example of high-performance envelope design which paid close attention to the existing environment, was part

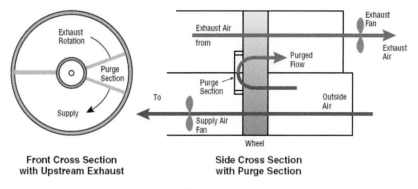

Front Cross Section
with Upstream Exhaust

Side Cross Section
with Purge Section

Figure 7-70

of a design-build process with multiple stakeholders involved, and used a broad portfolio of strategies to minimize energy transfers and provide flexibility for future improvements.

References
1. For a very clear and concise discussion of these properties, please see Mechanical and Electrical Equipment in Buildings, Fifth Edition, Grondzik, Kwok, Stein, Reynolds, pp. 183-187.
2. Please see Sources for more information.
3. A good definition of "biosphere" comes from http://www.businessdictionary. com/definition/biosphere.html: "Part of the Earth's surface and atmosphere that contains the entire terrestrial ecosystem, and extends from ocean depths to about six kilometers (3.7 miles) above sea level. Not precisely demarkable, it contains all living organisms and what supports them: soil, subsurface water, bodies of water, air and includes hydrosphere and lithosphere. Also called ecosphere."
4. Please SEE: http://www.webstatsdomain.com/domains/csep10.phys.utk.edu/
5. Please SEE: Autodesk/Ecotect: www.usa.autodesk.com/adsk/servlet/pc/index ?id=12602821&siteID=123112.
6. Mexican Hothouse at the Jardin des Plantes, Paris, 1836, (Jastrow, 2004).
7. Prudential Building. Louis Sullivan, 1894. Source: U.S. Library of Congress, Prints and Photographs Division, "Built in America" Collection
8. See: Chapter 3, Codes: IECC, for the climate map. Weather files for a particular location can be developed using Climate Consultant 5.1 available at: www. energy-design-tools.aud.ucla.edu/., or from EnergyPlus at: http://apps1.eere. energy.gov/buildings/energyplus/cfm/weather_data2.cfm.
9. Wind Roses can be simulated to predict wind conditions by using: www.enviroware.com/portfolio/windrose-pro/#axzz1gCPM7IqK.
10. Grid parity refers to the tipping point where the cost of supplying electricity from a renewable energy source becomes the same or less as the cost of supplying it from a fossil-fuel energy source.
11. It is important to make a distinction here with Number 3. The testing results

in a laboratory for the physical properties of building materials in an envelope do not always accurately predict the energy performance of the envelope in the field. This is one of the reasons that a National Database of High-Performance Buildings is being established to correlate predicted with actual energy performance.

12. For more expanded definitions of these properties, please see ASHRAE Handbook of Fundamentals (2007).
13. LEFT: http://www.syracuse.com/news/index.ssf/2009/08/carrier_to_lay_off_170_in_dewi.html,
    RIGHT: http://www.panoramio.com/photo/82284980
14. http://www1.eere.energy.gov/ba/pba/intensityindicators/efficiency_intensity.html
15. http://scienceworld.wolfram.com/physics/EnergyDensity.html
16. See the Buildings Energy Databook, http://buildingsdatabook.eren.doe.gov/, for definitions and information on Energy Intensity of the Buildings Sector (all sources).
17. In this section we will consider building materials related to heat transfer in the envelope exclusive of windows and glazing (covered in the next section: Light).
18. http://c.ymcdn.com/sites/www.polyiso.org/resource/resmgr/technical_bulletins/tb101_jun30.pdf
19. SHOWN IN ARCHIEXPO: http://www.archiexpo.com/prod/xtratherm/rigid-polyisocyanurate-and-plasterboard-insulation-panels-62174-263905.html
    MANUFACTURER: http://www.xtratherm.com/
20. http://coolroofs.org/
21. http://coolroofs.org/documents/ANSI:CRRC-1_Standard.pdf
22. http://www.astm.org/Standards/E903.htm
23. http://roofcoatings.org/
24. http://www.therrci.org/
25. http://www.ornl.gov/sci/roofs+walls/facts/CoolCalcEnergy.htm
26. http://www.roofcalc.com/
27. http://www.astm.org/Standards/E917.htm
28. http://dcgreenworks.org/resources/faq
29. http://www.astm.org/Standards/E2400.htm
30. http://www.igra-world.com/green_roof_literature/index.php
31. http://en.wikipedia.org/wiki/File:20080708_Chicago_City_Hall_Green_Roof.JPG
32. http://www.gmdconline.org/
33. http://www.gmdconline.org/index.php/buildings/810-humboldt-street
34. As quoted in Wikipedia: "A thermal bridge, also called a cold bridge,[1] is a fundamental of heat transfer where a penetration of the insulation layer by a highly conductive or noninsulating material takes place in the separation between the interior (or conditioned space) and exterior environments of a building assembly…" Allen, E., & Iano, J., Fundamentals of Building Construction:Materials and Methods. Hoboken, NJ: John Wiley & Sons Inc., 2009.
35. One from the DOE: http://basc.pnnl.gov/resource-guides/slab-edge-insulation
36. http://www.foamular.com/foam/
37. http://imgs.ebuild.com/guide/products/2005/2006/enpn/2008/0828/enpn.htm
38. http://www.asi-sd.com/products/foam-insulated-metal-roof-and-wall-panels.htm
39. Cradle to Cradle Certification: http://www.c2ccertified.org/

40. http://www.sips.org/technical-information-2/thermal-performance/sip-r-values/

41. http://www.sips.org/technical-information-2/thermal-performance/air-infiltration/

42. LITTLE BIGHORN HEALTH & WELLNESS CENTER: http://www.sips.org/green-building/bea/2012-building-excellence-award-winners/bea-2012-commercialindustrialinstitutional/little-big-horn-college-health-wellness-center/

43. LEFT: MMA Architectural Systems Ltd. http://7660.uk.all.biz/, RIGHT: Carl Stahl http://www.carlstahl-architektur.com/fileadmin/files/cs-arc/FacadeGreenerySystem_i-sys_en.pdf

44. http://buildblock.com/insulated_concrete_forms.asp

45. LEFT: Waterside Condominiums, Fort Myers, FL, in Structure Magazine: http://www.structuremag.org/article.aspx?articleid=353
RIGHT: 4000 sq. ft. Single-Family Home by Midwest Modern, LLC (http://www.midwestmodern.com/www.midwestmodern.com/Origins.html) using the Quadlock ICF System: http://www.quadlock.com/insulated-concrete-forms/news/111130_Value_of_Building_Insulation.htm

46. Building-Integrated Photovoltaic Designs for Commercial and Institutional Structures A Sourcebook for Architects, Patrina Eiffert, Ph.D., Gregory J. Kiss, p. 7, http://www.nrel.gov/docs/fy00osti/25272.pdf

47. http://www.sciencedirect.com/science/article/pii/S1364032105000055

48. http://www.sciencedirect.com/science/article/pii/S0038092X08002120

49. http://www.marketsandmarkets.com/Market-Reports/bipv-market-509.html

50. LEFT: CIS TOWER, Manchester, England, The Tower was retrofit with BIPV over the existing wall panels, three faces of the building were clad with a total of 7,244 Sharp photovoltaic panels generating 390kW of energy, http://en.wikipedia.org/wiki/Building-integrated_photovoltaics
RIGHT: SOLAR ROOF TILES, http://www.treehugger.com/renewable-energy/building-integrated-solar-power-tiles-now-available-with-sunrun-solar-as-service-program.html

51. http://en.wikipedia.org/wiki/File:Fa%C3%A7ana_Fotvoltaica_MNACTEC.JPG

52. For an excellent discussion of the issues surrounding moisture balance and thermal transmission in interior wall retrofits please see: http://www.building-science.com/documents/digests/bsd-114-interior-insulation-retrofits-of-load-bearing-masonry-walls-in-cold-climates

53. http://www.kraspan.ru/eng/?systems

54. 'Super-Insulated Shell' Called Key to Landmark Energy-Retrofit Project, in: http://www.durabilityanddesign.com/news/?fuseaction=view&id=5637

55. AS SHOWN IN: http://www.durabilityanddesign.com/news/?fuseaction=view&id=5637, copyright: www.buildingscience.com

56. PROJECT TEAM; Biome Studio, WinnDevelopment, Elton+Hampton Architects, Peterson Engineering Inc., and CWC Builders Inc.

57. http://www.eere.energy.gov/

58. http://www.expol.co.nz/expol-x.html

59. NOTE: This strategy could be employed with new construction or retrofit. http://www.ornl.gov/sci/buildingsfoundations/handbook/section3-1.shtml

60. http://www.lowenergyhouse.com/concrete-floor-insulation.html

61. Building Code inspections do not generally cover process requirements with installed components that insure adherence to manufacturers' specifications and validation of warranted performance outcomes in these areas. However, as we have seen in Chapter 6, there is increasing confluence between Building Code

authorities and High-Performance Standards.

62. SEE: ASTM E1186 - 03(2009) Standard Practices for Air Leakage Site Detection in Building Envelopes and Air Barrier Systems, http://www.astm.org/Standards/E1186.htm

63. SEE: Infiltration Modeling Guidelines for Commercial Building Energy Analysis, Pacific Northwest National Laboratory/DOE, 2009, http://www.energy.ca.gov/title24/2013standards/rulemaking/documents/public_comments/45-day/2012-05-15_Infiltration_Modeling_Guidelines_for_Commercial_Building_Energy_Analysis_TN-65229.pdf

64. Note: electrical lighting will be treated in Chapter 9: Mechanical & Electrical Equipment.

65. http://www.ies.org/explore/

66. http://windows.lbl.gov/daylighting/designguide/dlg.pdf

67. http://www.nfrc.org/

68. SEE: Daylighting Performance and Design, Ander, G., John Wiley, 2003.

69. http://continuingeducation.construction.com/article.php?L=120&C=423&P=3, IslandWood Environmental Center, Bainbridge Island, WA, Mithun Architects: http://mithun.com/

70. http://www.wbdg.org/resources/daylighting.php

71. http://elad.su-per-b.org/index.php?title=Section_Design

72. http://manhattan.about.com/od/artsandculture/ig/Guggenheim-Museum-Photos/Guggenheim-Skylight.htm

73. http://en.wikipedia.org/wiki/File:Galleria_Vittorio_Emanuele_II_(Milan)_E1.jpg

74. http://apps1.eere.energy.gov/buildings/publications/pdfs/commercial_initiative/toplighting_final_report.pdf

75. Solar Irradiance - the amount of solar energy that arrives at a specific area of a surface during a specific time interval (radiant flux density). A typical unit is W/m2. DEFINITION FROM: http://rredc.nrel.gov/solar/glossary/gloss_s.html

76. Here, the angle of the sun incident to the glazing plane.

77. A Testing Procedure for Conducting On-Site Evaluations of the Relative Solar Heat Gain Through Glazing Materials and Shading Attachments, Robinson, A.P., Griffith, W.E., FIES, Proceedings of the 19th National Passive Solar Conference, American Solar Energy Society, San Jose, CA, 1994. Article available from the author.

78. http://www.ppg.com/corporate/ideascapes/resglass/homeowners/product/Pages/ProdInfoSolarControl.aspx#azuria

79. U-Value of the Glazing System refers to the Total Value which includes the glass and the frame.

80. http://en.wikipedia.org/wiki/File:PlasmaCVD.PNG

81. http://www.efficientwindows.org/

82. http://www.csbr.umn.edu/

83. http://www.ase.org/

84. http://windows.lbl.gov/software/comfen/comfen.html

85. http://www.efficientwindows.org/gtypes_1.php

86. http://www.efficientwindows.org/gtypes_2lowe.php#lsg

87. http://www.ppg.com/corporate/ideascapes/SiteCollectionDocuments/Lo%20Res%2011597%20Solarban%2070XL%20(7097).pdf

88. LSG: the ratio of visible light transmittance to solar heat gain.

89. QUOTED IN: http://www.bdcnetwork.com/spectrally-selective-glazing%C2%A0more-daylight-less-heat-gain

90. http://www.cyberchromeusa.com/Color-QC-and-Matching-Blog/bid/26751/
    Color-Measurement-Accuracy-Translucent-Materials
91. LEFT: http://www.atlantechsystems.com/documents/translucent_glazing_sys-
    tems.html
    RIGHT: http://www.laceyglass.com/
92. CPI International, Inc. http://www.cpidaylighting.com/product.php/Quad-
    wall-Wall-Lights-70/
93. http://www.cpidaylighting.com/gallery-view.php/Building-Sectors-Daylight-
    ing-Retrofits-19/
94. Exterior fixed architectural treatments for shading were more common in the US
    in the 1940s – 1960s.
95. SEE: www.architecture.com, http://www.architecture.com/SustainabilityHub/
    Designstrategies/Fire/1-4-1-6-Solarcontrolsandshading.aspx
96. http://windows.lbl.gov/daylighting/designguide/section5.pdf
97. http://www.csbr.umn.edu/download/AMCA_fullreport.pdf
98. Ibid., p. 4
99. Ibid., p. 5
100. Ibid., p. 4
101. LEFT: http://www.bembook.ibpsa.us/index.php?title=Solar_Shading
     RIGHT: http://windows.lbl.gov/daylighting/designguide/section5.pdf
102. LEFT & RIGHT: Colt International Solar Shading. http://www.coltinfo.co.uk/
     downloads.html
103. http://peninsula.bcarc.com/technology.html#shading, DESIGN: Bercy Chen
     Studio
104. Construction Specialties. http://www.c-sgroup.com/solarmotion/external-
     venetian#node-2101
105. http://www.automatedbuildings.com/news/aug07/articles/
     zing/070723051101dilouie.htm
106. 3M Solar Control Films. http://multimedia.3m.com/mws/mediawebser
     ver?mwsId=SSSSSufSevTsZxtUNx_15x2xevUqevTSevTSevTSeSSSSSS—
     &fn=70-0709-0232-8.pdf
107. http://windows.lbl.gov/comm_perf/nyt_overview.html
108. http://radsite.lbl.gov/radiance/HOME.html
109. http://windows.lbl.gov/comm_perf/nyt_control-system.html
110. http://windows.lbl.gov/comm_perf/nyt_post-occupancy.html
111. ANSI S3.5-1997, Methods for Calculation of the Speech Intelligibility Index,
     American National Standard, Standards Secretariat, Acoustical Society of
     America, New York.
112. http://www.certainteed.com/silentfx/science-airborne.html
113. Reverberant noise control in rooms using sound absorbing materials., Warnock,
     A.C.C., Institute for Research in Construction, National Research Council of
     Canada, BRN 163, June 1980.
114. Sound isolation and fire resistance of assemblies with firestops, Nightingale,
     T.R.T., Sultan, M.A Construction Technology Update No. 16, Institute for Re-
     search in Construction, National Research Council of Canada, 1998.
115. http://bpcan.com/en-CA/products/insulation-and-structural-boards/compos-
     ite-insulation-panels/enermax-east-/
116. http://www.rdccanada.com/higherstandards.shtml
117. Acoustics in practice. Building Science Insight 1985, Warnock, A.C.C., Institute
     for Research in Construction, National Research Council of Canada, NRCC
     27844, pp. 39 – 51.

118. The COPE-Calc software available at http://irc.nrc-cnrc.gc.ca/ie/cope//eng/ ibp/irc/ctus/ctus-index.html can perform the speech privacy calculations as described in references 6 and 7.

119. Sound Propagation between Two Adjacent Rectangular Workstations in an Open-plan Office, I: Mathematical Modeling, Wang, C., Bradley, J.S., Applied Acoustics 63, (12) 1335-1352 (2002). DOI:10.1016/S0003-682X(02)00034-8, and: DOI:10.1016/S0003-682X(02)00035-X

120. http://ths.gardenweb.com/forums/load/oldhouse/msg051429468315.html

121. http://www.wbdg.org/references/casestudies.php

122. https://buildingdata.energy.gov/hpbd, and Zero-Energy Buildings: http://zeb.buildinggreen.com/

123. NOTE: When we say light mass we are not referring to the mass of the structure (which could be heavily massive), but only to the relative mass and thickness of the envelope (skin) itself.

124. SEE, e.g.: The Aldo Leopold Legacy Center, from the Zero Energy Buildings Database: http://zeb.buildinggreen.com/

125. http://en.wikipedia.org/wiki/Pantheon,_Rome

126. http://www.liechtensteinmuseum.at/en/pages/artbase_main. asp?module=browse&action=m_work&lang=en&sid=31032196&oid =W-147200412195342017

127. LEFT: The Bauhaus, Dessau, 1925, RIGHT: Wuhan, China, 2007, both: http:// en.wikipedia.org/wiki/Curtain_wall

128. http://eere.buildinggreen.com/mtxview.cfm?CFID=120324580&CFTOK EN=64734261

129. http://www.wbdg.org/references/casestudies.php

130. http://en.wikipedia.org/wiki/File:G2_Earthship_Taos_N.M.JPG

131. http://milieukontakt.mk/?p=1207

132. http://www.new-mexico-bed-and-breakfast.com/

133. http://www.aiatopten.org/node/115

134. http://blog.archpaper.com/wordpress/archives/tag/design-awards, PHOTO CREDIT: Aidlin Darling Design

135. http://www.aiatopten.org/node/115, PHOTO CREDIT: Aidlin Darling Design

136. http://eere.buildinggreen.com/overview.cfm?ProjectID=1156

137. http://building.cnt.org

138. http://building.cnt.org/

139. LEFT: East Elevation, RIGHT: West Elevation, PHOTO CREDIT: Timothy Hursley

140. http://eere.buildinggreen.com/process.cfm?ProjectID=286

141. http://eere.buildinggreen.com/energy.cfm?ProjectID=286

142. CREDIT: Croxton Collaborative Architects, http://eere.buildinggreen.com/images.cfm?ProjectID=286

143. "Energy recovery ventilation (ERV) is the energy recovery process of exchanging the energy contained in normally exhausted building or space air and using it to treat (precondition) the incoming outdoor ventilation air in residential and commercial HVAC systems. During the warmer seasons, the system pre-cools and dehumidifies while humidifying and pre-heating in the cooler seasons." Dieckmann, John. "Improving Humidity Control with Energy Recovery Ventilation." ASHRAE Journal. 50, no. 8, (2008), as quoted in: http://en.wikipedia.org/wiki/Energy_recovery_ventilation

144. http://eere.buildinggreen.com/energy.cfm?ProjectID=286

Chapter 8

On-site Distributed Power (DEG)

Paul D. Tinari Ph.D., P.Eng.

INTRODUCTION

The electric power sector is presently facing three energy-related challenges—environmental sustainability, security of supply, and competitiveness—all within a context of growing electricity demand. These issues may also represent drivers for the further penetration of distributed power generation (DPG)[1] technologies. Grid reliability and the rapid growth of the green building sector has been the major trend driving the steadily increasing popularity of DPG systems in energy efficient buildings.

This trend is also driven by rapidly emerging technologies that have permitted the development of cheaper, more efficient and smaller generating units. The traditional approach of adding new large, central generation and transmission capacity is now increasingly frustrated by social, economic and environmental constraints. These impediments are contributing to the increasing popularity of DPG systems.

Achieving a comprehensive, sustainable, interconnected, intelligent, national energy system requires extensive energy infrastructure changes, which represent one of the most important investments of the 21st century. The present chapter focuses on the potential role of DPG systems in high-performance buildings. For the purposes of this chapter, DPG is defined as an electric power source connected to the power distribution network, serving a customer on-site or providing network support. DPG technologies may consist of small or medium size, modular energy conversion units, which are generally located close to end users and transform primary energy resources into electricity (and in some cases provide heat and cooling via combined heat and power (CHP) technologies).

The most important advantages inherent to the DPG concept include:

- *Increasing the Security and Reliability of Supply*: By adding on-site generating capabilities.

- *Advantages in Transmission and Distribution Network Operation and Planning*: Distribution networks with high DPG penetration reduce overloading of the utility transmission grid. On-site DPG systems greatly reduce transmission and distribution losses.

- *More Efficient use of Primary Energy Resources and Waste Heat*: DPG units enable more efficient exploitation of available energy sources like waste heat, biomass and renewable energy sources.

- *Reduction of Noxious Pollutant Emissions*: Renewable energy sources and/or efficient DPG systems can significantly reduce fossil fuel consumption and greenhouse gas (GHG) emissions.

- *Energy Market Competitiveness*: DPG systems can stimulate competition in permitting more generators to supply the electricity market.

- *Authorization and Construction Facilitation*: It is easier to find sites for and get permits for DPG systems than for large central power plants and DPG systems can be brought on-line much more rapidly.

However, there are also a number of challenges associated with the installation of DPG systems in buildings and with their integration into the utility power distribution networks. Some of the most important of these issues include:

- *Distribution Planning and Operation*: Local electrical distribution systems may need to be altered or redesigned since DPG technologies may significantly alter local power flow patterns.

- *Data Collection*: Accurate data are needed for controlling the DPG system and this can sometimes be difficult to gather.

- *Power Reserve and Balancing*: Due to the intermittent nature of the output from many DPG sources, the automatic control system must be able to deal with transients in local power generation and when necessary, be able to tap into the needed power reserve from the external utility transmission system. When the local power system includes intermittently available renewable energy sources, it becomes increasing difficult to balance the entire system when confronted with rapidly varying building loads.

- *Occasional Lower Power Quality*: Harmonics may be generated by electronic power converters used to connect some DPG systems.

- *Potential Adverse Impacts on the Grid*: The effects of DPG developments on the regional and even the national transmission system cannot be neglected. Without a properly coordinated system interface and flexible controlling devices, the consequences of a power disruption at the distribution level may be suffered (if not amplified) at the transmission level.

The fact is that more efficient and environmentally friendly energy generation technologies have continued to emerge in the fields of cogeneration (combined heat and power or CHP), renewable energy sources (RES), fuel cells and many other areas. In addition, steady technological progress has been made in the utilization of power electronics-based inverters, rectifiers and other converters. These devices play a vital role in the connection of small/medium power units to the grid.

The field of DPG systems is potentially vast and detailed reference works are available that explore the topic in significant detail. For this reason, this chapter will focus only on a general description of CHP and RES.

DPG SYSTEMS TECHNOLOGIES AND APPLICATIONS

Selected DPG technologies have been chosen for closer examination in this chapter due to their potential role with respect to the objectives of security of supply, competitiveness and emissions reduction.

DPG systems technologies range from non-renewable (like internal combustion and Stirling engines, combustion turbines, microturbines and fuel cells) to renewable energy source technologies (like wind turbines, small/micro hydroelectric plants, photovoltaic and solar thermal units, biomass units and geothermal plants)[2]. DPG technologies can be classified according to their different properties and features, for example, depending on their suitability for CHP production. Energy storage technologies (such as batteries and flywheels), whose utilization can be an option for managing intermittency of some DPG technologies, are still very much in development as far as stand-alone buildings are concerned, and are beyond the scope of this chapter.[3]

The first important factor to consider in deciding whether generating units can be inserted into a particular distribution network, is what the maximum net power rating is at the connection point. A second factor is the operational flexibility of the units. The flexibility of DPG system operation may be defined as the ability of units to respond to a change in the demand within an acceptable time frame. It is also possible that the DPG system may generate power when it is not required and the control system must be able to feed excess power into the regional utility grid. At certain times, intermittent DPG electrical output may not produce enough power to meet building requirements and the control system has to be able draw the necessary power from the grid.

One of the most significant features of DPG technologies is their modularity. DPG system units generally come in standard sizes and can be easily combined with other compatible units to produce a higher capacity system.

There are many advantages associated with DPG system modularity:

- *Adaptability*: It is possible to generate a broad spectrum of possible power outputs simply by modifying the number of modules.

- *Increased System Reliability*: The failure of one module will not necessarily imperil the operation of the rest of the power generation system.

- *Shorter Commissioning Time*: Modules are generally available off the shelf and conform to industry wide standards, reducing the procurement and installation time.

- *Reduced Costs*: Modular "mix and match" systems are much cheaper to procure, install, operate and maintain.

COGENERATION

When there is a significant and relatively constant demand for process heat, DPG systems that are suitable for CHP generation may be economically attractive. In such cases, use of CHP systems may be a much better choice than simply purchasing electricity from the grid and producing the required heat separately. Presently, industries that need steam for processes in addition to electrical power may benefit from this CHP feature. As the costs of CHP fall, it is only a matter of time before households benefit from these technologies for the supply of hot water for space heating and for the generation of household electrical power.

Cogeneration can greatly increase the overall (heat and electricity combined) system efficiency to as high as 90% depending on DPG type and size[4]. Figure 8-1 provides an example of the difference in efficiency between separate and combined heat and power generation. The CHP system used in the example can convert 85 out of 100 units of input fuel into useful energy (45 heat, 40 electricity). On the other hand, to have the same amount of heat and power generated by separate heat (in a boiler) and power (in a plant) generation, 160 units of input fuel would be necessary.

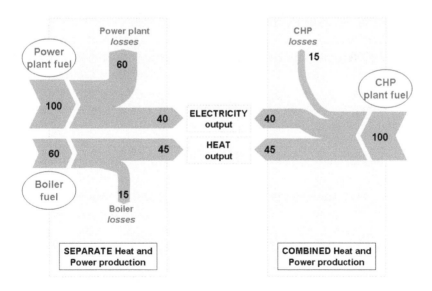

Figure 8-1

However, one consequence of CHP deployment is that the facility's electrical power supply becomes dependent on heat demand. This aspect has to be taken into account at the planning and design stage and may impact the ability of a CHP DPG system to follow demand changes.

A list of appropriate technologies suitable for CHP would include internal combustion engines, industrial combustion turbines, microturbines, biomass gasification units, geothermal units, Stirling engines and fuel cells.

ENGINES

The predominant DPG systems technology is currently an internal combustion engine (ICEs) driving an electric generator. ICEs have gained widespread acceptance in many sectors of the economy, serving in most cases as backup generators for sensitive loads such as hospitals, for which long-duration energy supply failures would have serious consequences. ICEs are currently available from numerous manufacturers in all imaginable DPG sizes ranging from a few kW to 50 MW or more. Smaller engines are primarily designed for transportation applications and can be converted to building power generation with little modification. Larger engines are, in general, designed for power generation, mechanical drive, or marine propulsion.

Reciprocating engines can be either spark ignition (Otto cycle) or compression ignition (Diesel cycle). When properly adapted, spark ignition engines can use a wide variety of fuels including gasoline, natural gas and biogas from a variety of sources including sewage plants. Diesels can be modified to run on bio-diesel fuels and at least one company has managed to design an adaptor that allows them to run on natural gas.[5]

The efficiency of most commercial ICEs ranges from 30-38%, reaching a maximum of more than 45% for advanced diesel engines. Most ICEs rotate at between 1500 or 3000 rpm and can easily be coupled to either four-pole or two-pole electric generators and can be exactly controlled allowing them to be coupled to the grid. Diesels can be put on line extremely quickly and they can quickly adapt to rapidly varying loads, which is why they are the staple of organizational emergency power supplies. They can be used to drive synchronous generators at 750, 1500 or 3000 rpm.

One of the disadvantages with using ICE in DGP systems is that they are fairly large and they usually require a significant footprint to be set aside to mount the engine-generator set, control and water cooling system. It is relatively easy to incorporate ICEs into cogeneration systems as significant amounts of lower grade heat is available from the engine's cooling jacket and lubricating oil. High grade heat can be recovered from the engine's exhaust.

Figure 8-2 shows a typical CHP system that could be used in a building such as a hospital.

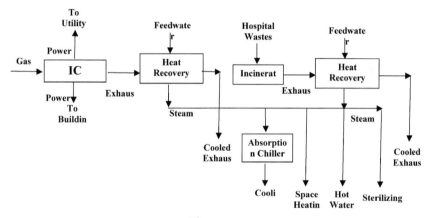

Figure 8-2

TURBINES

Turbines come in many types and can be driven by a variety of different fuels. A gas turbine has one section that compresses ambient air and combines it with the gaseous or liquid fuel. The ignition of the air/fuel mixture drives the turbine as the combustion products are expanded through the blades. In more advanced models, the compressor and axial bladed turbine can be equipped with multiple stages to increase the efficiency of the system. Radial, single stage microturbines that have an output of a few kilowatts are now available for various green building applications.

The advantages of turbines include low installation costs and emissions along with the easy recovery of heat using steam or oil and with relatively low maintenance requirements. However, system efficiencies can be rather low, ranging from 25-30%. Turbines are generally used for cogeneration in a DPG system when a continuous supply of steam, hot water and power is needed.

Turbine-based Cogeneration Systems

There are two approaches in general use, namely, using back pressure or condensing turbines and extraction turbines. In the first of these, all of the flow through the turbine is used to operate the cogeneration system. The exhausted steam is condensed at a relatively high pressure and temperature and the heat is then recovered for use within the building. The disadvantage is that the electrical efficiency is relatively low and when there is little demand for the waste heat, it is uneconomical to operate the turbine. In a cogeneration system using extraction turbines, a portion of the steam flow is drawn off, condensed and the heat recovered. The remaining steam is allowed to flow down the axis of the turbine to exit at a low pressure before it is condensed. There are a number of advantages with this approach including the fact that in periods of low heat demand the extraction system can be closed off, maximizing electrical output. Also, since the condenser pressure is low, the electrical efficiency tends to be higher than with back pressure systems.

The main advantage of steam turbine based cogeneration systems is that, because the steam is generated in a boiler, virtually any type of fuel can be used. Systems can be designed to use waste biomass, natural gas, bunker C oil, low grade fuels or even concentrated solar energy.

Steam turbines for DPG systems are able to drive synchronous generators, although a gearbox may be necessary to reduce the rotational velocity to an acceptable level. Both types of turbine can be made to respond to small changes in demand by use of a control system that limits the mass flow rate of steam through the turbine. However, larger oscillations in demand have to be handled by modifying the rate of steam production from the boiler. Another limiting factor will be the ability of the cogeneration system to deal with changes in the flow of steam. It is possible for the extraction turbine system to direct all of the steam flow through the condenser to adapt to the electrical requirements placed on the system.

Gas Turbines

Modern gas turbines are constructed of materials that do not require cooling and they can have efficiencies of more than 40%. Their main disadvantage is that they require high quality fuels such as clean natural gas or high purity liquid fuels. They also become significantly less efficient at low power outputs. For this and a number of other reasons, the most appropriate use of a gas turbine for DPG systems is to couple it directly to a synchronous generator.

The design of a typical CHP system using a gas turbine is shown in Figure 8-3.

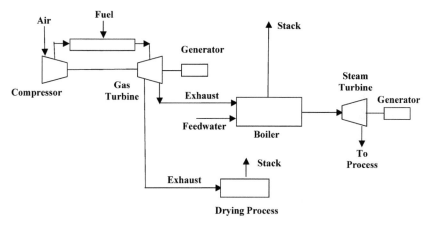

Figure 8-3

Microturbines

Microturbines are a type of combustion turbine that can be easily harnessed to produce both heat and electricity on small scales. They offer a number of advantages including having a relatively small number of moving parts, small size and weight, high efficiency, low emissions, low cost and the ability to use a wide variety of fuels. When integrated into a well designed cogeneration system, the overall efficiency can exceed 80%.

Microturbines can be designed to produce electrical outputs ranging from 1 kW to more than 500 kW. The first microturbines were built to serve as vehicle turbochargers or as auxiliary power units (APUs) for various applications. The performance of modern microturbines has been significantly improved by the use of advanced high temperature and high strength ceramics, and with the use of barrier coatings. Each microturbine is made up of a turbine, compressor, combustor, alternator, recuperator and a generator. These small combustion turbines can be single shaft or two-shaft, simple cycle, or recuperated, intercooled and with reheat. Because of their greater simplicity and lower cost, single shaft microturbines rotating at over 100,000 RPM are more commonly selected.

Some of the highly desirable characteristics of microturbines include:

- *Ideal for DPG Systems*: They can operate autonomously and unattended for long periods providing relatively low-cost power.

- *Producing High Quality Power*: Systems produce power with only minor variations in frequency and voltage.

- *Instantly Available*: Able to provide standby or peak reduction power.

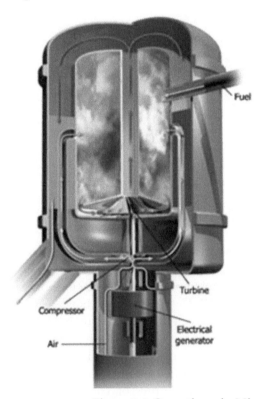

Fuel enters the combustion chamber. The turbine can run on natural gas, gasoline, kerosene— virtually anything that burns.

Fuel

The hot combustion gases spin a turbine, which is connected to the shaft of an electrical generator. The exhaust transfers heat to incoming air.

Turbine

Compressor

Electrical generator

Air

Air passes through a compressor and is warmed by the exhaust gases before entering the combustion chamber.

Figure 8-4. Operation of a Microturbine[6]

There are two general types of microturbine systems:

- *Simple Cycle*: Compressed air is mixed with the fuel and then burned under constant pressure before the hot exhaust is expanded through the turbine. These systems have lower efficiencies but offer high reliability, lower costs and a greater amount of heat available for operation of cogeneration systems.

- *Recuperated Cycles*: These systems use heat exchangers to recover energy from the exhaust and then use it preheat the incoming air stream before it enters the combustor. More energy is recovered for use in the cogeneration system. The use of the recuperator can reduce fuel use by up to 40%. Figure 8-5 is a diagram of such a system. The overall efficiency of a microturbine combined with a cogeneration system can exceed 85%.

Modern microturbines are available at costs below $1000/kW with some low cost units going for below $650/kW. Because the

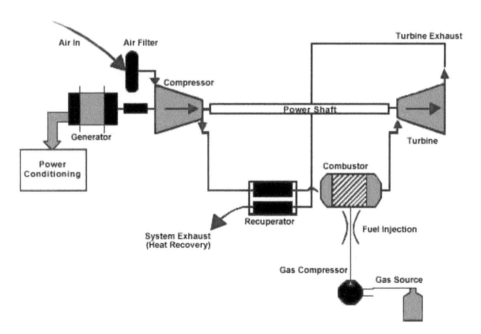

Figure 8-5. A Recuperated Microturbine System[7]

units have so few moving parts and are based on robust and proven designs, they offer high reliability, requiring maintenance only every 6,000 to 8,000 hours. Operation and maintenance costs will range from $0.005 to $0.017/kW (source: California Distributed Energy Resources Guide).

Electrochemical Energy Conversion Devices (Fuel Cells)

The fuel cell is a surprising old technology. The first operational fuel cell was developed by Sir William Grove in 1839. Grove observed that water could be split into hydrogen and oxygen by sending an electric current through it and so he hypothesized that by reversing the procedure, electricity and water could be produced. The term fuel cell was coined about fifty years later by scientist Ludwig Mond.

There are four basic components of most fuel cells:

- *Anode*: Conducts the electrons that are freed from the fuel molecules so that they can be used in an external circuit. It often has channels etched into it that disperse the incoming fuel equally over the surface of the catalyst.

- *Cathode*: Has channels etched into it that distribute the oxidizer to the surface of the catalyst. It is also responsible for conducting the electrons back from the external circuit to the catalyst, where they can recombine with the fuel ions and oxidizer.

- *Electrolyte*: This is where the proton exchange takes place. This is a material that is specially designed to conduct positively charged ions while blocking the flow of electrons. Cells have been built with membranes made of materials such as perfluorosulfonic acid and more recently, various aromatic-based structures.

- *Catalyst*: A material that facilitates the reaction of oxidizer and fuel. It can be made of coated nanoparticles incorporated into a thin layer spread on a membrane. The catalyst is rough and porous to maximize the surface area that can be exposed to the fuel and oxidizer.

The chemistry of a typical fuel cell is as follows:

Anode side:

$$2H_2 \Rightarrow 4H^+ + 4e^-$$

Cathode side:

$$O_2 + 4H^+ + 4e^- \Rightarrow 2H_2O$$

Net reaction:

$$2H_2 + O_2 \Rightarrow 2H_2O$$

This reaction in a single fuel cell produces a relatively low voltage, usually less than 1 volt. To raise the voltage up to a useful level, many separate fuel cells must be combined to form a fuel-cell stack. While long-term stability is still a problem because of effects such as corrosion, bipolar plates are commonly used to connect one fuel cell to another and are subjected to both oxidizing and reducing conditions and potentials. Low-temperature fuel cells commonly use lightweight metals, graphite and carbon/thermoset composites as the bipolar plate material. If the fuel source of a particular fuel cell is not pure hydrogen, then a reformer will be needed to convert the hydrocarbon or alcohol based fuels into hydrogen. This process will generate heat and produce other gases in addition to hydrogen. It is possible to use various methods to clean up the hydrogen, but even so, the hydrogen that emerges is not pure, and this has the effect of lowering the efficiency of the fuel cell. Because reformers impact fuel cell efficiency, researches have generally decided to concentrate on pure hydrogen fuel cells, despite the significant challenges associated with hydrogen production and storage. Figure 8-6 is a diagram of a typical fuel cell.[8]

There are many types of fuel cells commonly classified by operating temperature and electrolyte used. The principal types include:

- *Polymer Exchange Membrane (PEM)*—These have high power densities and low operating temperatures (60-80°C) which makes them ideally suited for transportation applications.

- *Solid Oxide (SO)*—These operate at very high temperatures (700-1,000°C) and are well suited for large-scale stationary power applications. While they are very stable when in continuous use, they can suffer from mechanical failure when subjected to thermal cycling. Because of the extremely high operating temperatures, this type of full cell is ideal for use in CHP systems.

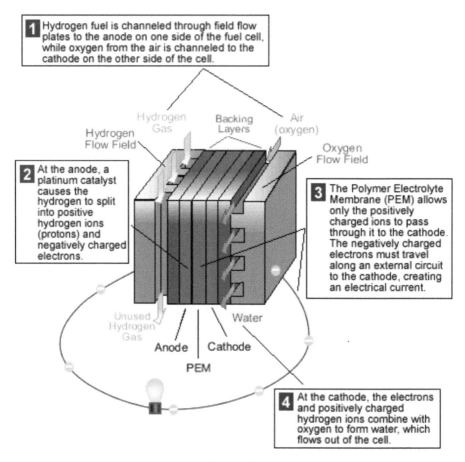

1 Hydrogen fuel is channeled through field flow plates to the anode on one side of the fuel cell, while oxygen from the air is channeled to the cathode on the other side of the cell.

2 At the anode, a platinum catalyst causes the hydrogen to split into positive hydrogen ions (protons) and negatively charged electrons.

3 The Polymer Electrolyte Membrane (PEM) allows only the positively charged ions to pass through it to the cathode. The negatively charged electrons must travel along an external circuit to the cathode, creating an electrical current.

4 At the cathode, the electrons and positively charged hydrogen ions combine with oxygen to form water, which flows out of the cell.

Figure 8-6

- *Alkaline*—This type of fuel cell has been in use by the space program since the 1960s. The main limitation is that their performance can be undermined by exposure to contamination, so they must be supplied with absolutely pure fuel. Another problem is their high capital costs, so they are not an ideal choice for most commercial applications.

- *Molten-carbonate (MC)*—Like the SO cells, these are also best suited for large stationary power generators. They operate at more than 600°C, so they can be used to create steam that can be used to generate additional power. They have a lower operating temperature than SO fuel cells, which means they can be built from cheaper, lower temperature materials.

- *Phosphoric-acid (PA)*—PA fuel cells has the potential to be used in small stationary power generation systems. They operate at a higher temperature than PEM cells, so they have a longer warm-up time. This makes them unsuitable for use for most transportation applications.

- *Direct-methanol (DM)*—While operating at similar temperatures, these cells are less efficient than PEM cells. Also, DM cells require a relatively large amount of catalyst, making the cells to expensive for many applications.

The main advantage with fuel cells is that they produce no pollution. Other advantages include their high efficiencies ranging from 50-60% and that fact that they have no moving parts, meaning that they generally display high reliability. Also, because some cells operate at high temperatures, they can offer many opportunities for the development of CHP systems and reducing the environmental impacts of green buildings.

Along with their significant advantages, there are a few major issues associated with fuel cells:

- *Cost*: Most of the component pieces of high quality fuel cells are expensive. Currently (2012), the projected high-volume production price is about $43 per kilowatt.[9] New lower-cost catalysts are needed to make fuel cells more affordable.

- *Durability*: It is imperative that new membranes be developed that have longer operational lives and that can operate at temperatures greater than 100°C while at the same time being able to function when ambient temperatures are below zero. In general, the higher the standard operating temperature of the cell, the greater the tolerance will be for impurities in fuel. Membranes will also tend to degrade faster when exposed to thermal cycling conditions. This makes them more suitable for stationary applications such as in green buildings, instead of for transportation uses.

- *Hydration*: Membranes must by hydrated to transfer hydrogen protons. It is necessary to develop fuel cell systems that can operate in low temperature & in low humidity environments. At the same time, for many membranes that are operated at high temperatures, hydration is lost without the use of a high-pressure hydration system.

Research has been ongoing at the Pacific Institute for Advanced Study on advanced, high-efficiency fuel cell systems.[10] Many of the latest fuel cells now used in commercial applications incorporate a number of complex, high cost subsystems. For example, the thermal management subsystem of some high temperature fuel cells includes a large radiator, heat exchanger, cooling plates, an energy management system (e.g. a contact cooler) and associated pumps and coolant controllers. Its purpose is to stabilize the temperature within the fuel cell stacks.

The water management subsystem includes reservoirs, accumulators, humidifiers, condensers and associated pumps and controllers. Problems arise in the fuel cell when too much liquid water is created. The tiny channels that the hydrogen and oxygen gases need to flow in can be blocked by water condensate. Simply increasing the flow of air through the system would cool the cell but this would also dehydrate the membrane. The membrane must remain moist for protons to pass through it. Research is focusing on integrating the water and the thermal management subsystems to simplify the manufacturing process and to reduce system costs.

For intermittent applications, an ultracapacitor can be used to recover and store energy that can used to assist in the startup of the system after a shutdown.

The blower supplies air at the appropriate pressure and flow rate to the fuel cell stacks. The air system often uses off-the-shelf compressors or blowers that are not specifically designed for fuel cell applications, resulting in a systems that is heavy, costly and inefficient.

These subsystems significantly increase the complexity and cost of the fuel cell system while reducing overall efficiency. The key to augmenting the commercial acceptability of fuel cells is to implement a paradigm shift in the design that achieves simplicity by eliminating most of these complex subsystems. A unique and innovative solution is to replace all of these components with a single integral fuel cell/flywheel system. A flywheel rotating at high rpm acts as an efficient energy capturing & storage system, eliminating the need for ultracapacitors. The flywheel also acts as a turbine gas compressor, eliminating the need for an air blower. The high centrifugal forces within the flywheel drive the flow of gases within the membrane, increasing reaction rates, while at the same time pulling water vapor and any liquid droplets that may have formed in the membrane, preventing channel clogging and dry-out. The flywheel also eliminates the need for the thermal management system since the rapidly rotating disk is able to efficiently dissipate extremely high heat fluxes. By eliminating most of the subsystems used in traditional fuel cell vehicles, it can be estimated that the new system will significantly reduce overall system costs. A flywheel schematic is shown in Figure 8-7.[11]

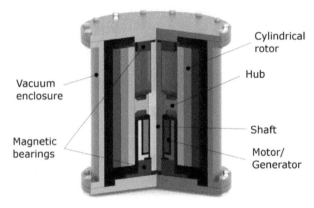

Figure 8-7

Reliability will also be significantly increased because the new system will only have one moving part—the flywheel—a well known, reliable and fully tested technology. The innovation is incorporating a fully operational fuel cell into the structure of the flywheel. The proposed design is shown in Figure 8-8.

STIRLING ENGINES

The principal idea of the Stirling engine is that a fixed amount of a gas is sealed inside the cylinder. The cycle involves changing the pressure of the gas in the cylinder causing it to do work. These have been proposed for the production of electricity in building CHP systems. The big advantage with Stirling engines is that they are significantly less complex than other reciprocating engines. They differ

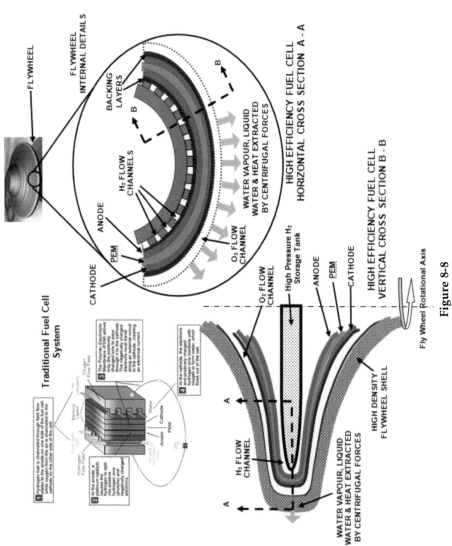

Figure 8-8

from IC engines in that they use external combustion. A cycle is created by using the difference between the hot and cold end of a fixed amount of gas expanding and contracting within the engine. The heat is generated externally by the combustion of a fuel-air mixture that can be accurately calibrated. Consequently, the creation of potential air pollutants such as NO_x can be minimized. This type of engine will operate equally well on virtually any type of fuel so it holds enormous potential as a device that can convert any alternate energy source including solar, geothermal, biomass etc. into electrical power. There are a number of possible configurations of Stirling engines depending on the heat source used. Modern Stirling engines are available in sizes ranging from 1 kW up to more than 100 kW.

Figure 8-9 shows a design of a simple Stirling engine.

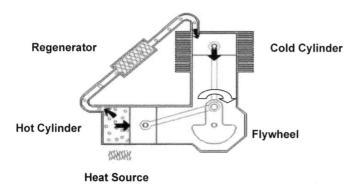

Figure 8-9

This is the original design produced by Robert Stirling in 1816 who was attempting to develop a very simple engine that was safer than the steam engines of the day that were prone to boiler explosions. In this design, the rotation of the flywheel is clockwise. The steps involved with the operation of this engine are as follows:

- *Expansion*: The gas is mostly in the "hot" cylinder. The heat input causes the gas to expand, driving both pistons inwards.

- *Transfer*: The hot gas expands, driving the hot piston to its maximum displacement to the right. The momentum of the flywheel starts to drive the hot piston to the left and the cold piston downwards. Most of the gas is transferred to the cold cylinder.

- *Contraction*: Most of the expanded gas has been transferred to the cold cylinder where it cools and contracts, causing both pistons to move outwards.

- *Transfer*: The compressed gas is still mostly in the cold cylinder and the momentum of the flywheel continues the motion of the cold piston upwards so that the gas is transferred back to the hot cylinder to complete the cycle.

The regenerator consists of high surface area matrix with high heat conductivity. When hot gas flows to the cold cylinder, it must first transfer some of its heat into the material of the regenerator. When the cooled gas flows back, this stored heat is retransferred to the gas. The regenerator therefore acts to pre-heat and pre-cool the working gas, dramatically improving efficiency.[12] Some of the advantages of Stirling engines include:

- Much higher efficiencies are possible than typical IC engines.

- The gas used inside the engine never escapes. There are no exhaust valves and there are no air/fuel explosions. Because of this, the operation is very quiet, making them ideal for use in green building applications.

- The Stirling cycle uses an external heat source, which could be fueled by anything from gasoline, solar energy and other low-grade, lower cost fuels to the heat produced by decaying plants No combustion takes place inside the cylinders of the engine.

There are a number of reasons why Stirling engines are not more popular for many applications, including in most commercial transportation. The use of an external heat source means that the system has significant hysteresis so the engine requires appreciable warm up time before it can begin operations and it cannot rapidly adapt to changing external power demands. A diagram of a 55 kW Stirling cycle engine that could have green building applications is shown in Figure 8-10.

SOLAR ENERGY

The two major approaches to harnessing incident solar energy are with photovoltaic (PV) and solar thermal (ST) collection systems. PV

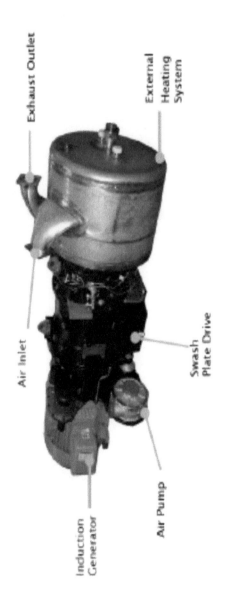

Figure 8-10

systems consist of arrays of modules made up of discrete interconnected semiconductor cells that convert solar radiation into electrical power. The PV cells generate DC power that may be converted to AC by use of an inverter. There are two main types of PV systems: off-grid and grid-connected. Off-grid systems are commonly used to supply electrical power in remote areas where there is no utility grid-supplied power available. Figure 8-11 shows the basic components of a grid-connected system.[13]

There are numerous advantages associated with the use of PV in DPG systems:

- Increasing affordability, with prices falling steadily in the last 10 years.

- Modularity, so the system can easily be expanded as energy needs grow or as budgets allow as long as the initial installation employs larger gauge wires, switching gears and controls.

- Minimal on-going operational costs.

- Emission-free operation.

- Durability, with some systems lasting for 20 years or more.

- Minimal maintenance and high reliability, since there are no moving parts.

With steadily increasing energy costs, and with the desire to reduce the impacts that buildings have traditionally had on the environment, the demand for PV systems is expected to continue improving.[14] PV systems consist of the following components:

- *PV Module*: A number of interconnected solar cells that are then assembled into panels and arrays.

- *Mounting Brackets*: Structures that attach the PV module to a roof or onto the ground.

- *Sun Tracking System*: Advanced PV installations will have a motorized system for changing the orientation of the panels so that they are always perpendicular to the sun throughout the day.

- *Electronics, Controls and Metering*: The DC power produced by the

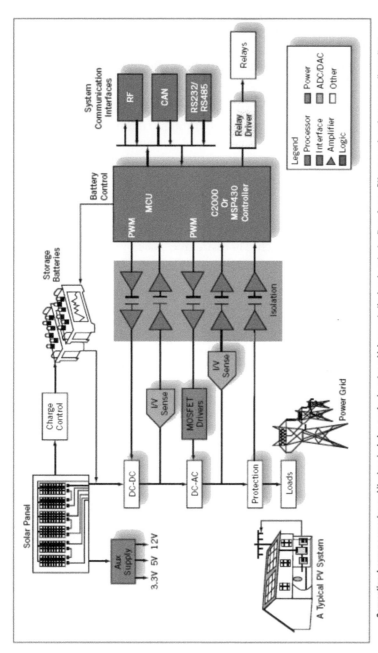

Conventional power converter architecture includes a solar inverter, which accepts the low dc output voltage from a PV array and produces an ac line voltage.

Figure 8-11

PV panels can be converted into AC power using inverters. If the panels are connected to the grid, then the amount of power coming into the building to make up for the deficiency in the solar generated power will be metered. If more power is generated by the PV system than is needed by the building, then the meter will measure the amount of power "sold" to the utility.

• *Backup Power*: when the sun is not shining, some form of backup power is required to supply the building. If it is not desired or possible to use utility power from the grid, then another alternate power source can be used such as wind turbine or microturbine power.

Configuration of the System for PV Net Metering

It is highly desirable to integrate PV systems into new building construction, but existing buildings can be retrofit efficiently with an integrated systems approach. There are five basic requirements to make a building "solar ready":

1. Selecting and/or designing a roof section of suitable size, pitch and orientation;

2. Installing ducts or conduits from the mechanical room to the roof area directly below the area where the future PV system will be installed;

Figure 8-12[15]

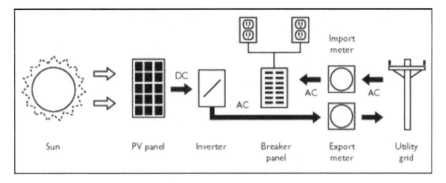

Figure 8-13

3. Including additional plumbing valves and fittings on the water heater if it is desired to install a future solar hot water system;

4. Provision of electrical outlets to power monitoring equipment and/or water circulation pumps; and

5. Instructions to the designer that the blueprints of the building indicate the location of all future alternative energy components.

Whenever it is desired to provide solar energy input to any building, it is important, if possible, to orient the structure on the lot to maximize its solar exposure. Wires with sufficient gauge and wire conduits with adequate capacity must be installed before the interior walls are enclosed. As solar systems generate low-voltage DC power, unless inverters are included at panel locations, the system wires will normally be larger than normal house wiring. Therefore to keep costs low, the length of wire runs through the structure should be minimized by careful attention to design beforehand. Some panel manufacturers are including micro-inverters at each panel and power optimizers to reduce line losses and improve energy harvesting by 5-10% to handle these issues.[16] Some utilities have been moving to "time-of-day" billing, which is highly favorable for buildings equipped with net-metered PV systems. Solar systems can generate excess electricity during periods of maximum sunshine, which also happens to correspond to peak electrical energy consumption times, and this electricity can be sold back to the grid in some jurisdictions under net-metering provisions. Two standard types of solar cells used in arrays are as follows.[17]

Crystalline Silicon(CS)PV   Peak power: 11.1 to 16.7 W/Sq Ft
CIS/CIGS, Thin Film   Peak power: 8.4 to 12.0 W/Sq Ft

As a quick, general example application, let's take a 100' x 400' warehouse building with an unshaded, flat roof. Let's say 80% of the roof area could be covered with a PV array. That's roughly 32,000 sq. ft. of panels. Take CS-PV for the example, with an average conversion at peak power of 14 W/sq. ft., giving us 448,000W or 448kW of production capacity. A very useful program from NERL, PVWatts2[19], can be used to calculate the annual energy production in kWh. A few examples, using an energy conversion factor of .8 and a PV Module tilt angle the same as the latitude angle with adequate separation between rows of modules to prevent shading and facing an azimuth of 180 degrees:

Location	Zip Code	kWh/Yr
Phoenix	85001	752,918
Dallas	75201	653,770
Pittsburg	15211	512,146

Depending on the usage in the building and the occupant's electrical requirements, this is substantial power production. In addition, 80% of the roof is not subjected to direct solar radiation which also reduces heat transfer into the building during cooling degree days.

Complete PV systems are steadily falling in price per installed kilowatt.[20] The next generation of PV technologies will offer both dramatically increased efficiencies along with significantly reduced costs. The long-term goal is to achieve total installed costs of less than $1000/kW. A number of new PV products have now become commercially available such as a flexible solar "shingle" that is available in the same size and shape as traditional asphalt shingles. The PV shingle enables the roofs of residential or commercial buildings to serve two purposes at the same time, namely, to be a source of power for the underlying building as well as providing moisture protection. That's more aggregate utilization. See Figure 8-14.

Another technological development

Figure 8-14

in solar-electric is using quantum dots. Quantum dots are particles of semiconductor material that are so small that electrons within them can only exist at certain specific energies.[21] These electron energy levels are exactly defined by the size of quantum dots, and so in turn, each quantum dot specifically defines the band gap, or the energy of the incident photon that will be absorbed. Using techniques developed by the semiconductor industry, the dots can be grown to any desired size, allowing photons of virtually any energy level to be absorbed without modifying the background material or changing construction techniques.[22] It is theorized that commercially available quantum dot PV cells could eventually offer efficiencies exceeding 80%, more than three times higher than the efficiencies of conventional systems available today. See Figure 8-15.

Solar thermal systems collect incident solar radiation and convert it into usable heat energy. In this application, solar energy can be collected using various types of flat panels or by using concentrating systems. The latter can generate significantly higher temperatures that are suitable for electrical power generation using small organic-cycle Rankin turbines. The relatively low temperatures produced by flat panels can also be used to power a number of useful applications such as driving absorption air conditioning units. Figure 8-16 shows a high efficiency "tube" type solar thermal panel.

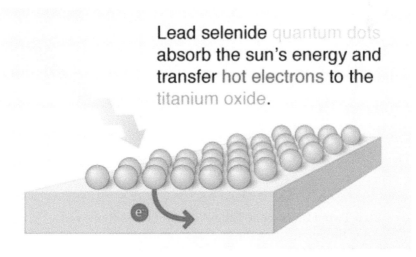

Lead selenide quantum dots **absorb the sun's energy and transfer** hot electrons **to the** titanium oxide.

Figure 8-15

Figure 8-16

While being significantly more efficient than traditional flat panels, tube collectors are also much more expensive. The Pacific Institute for Advanced Study (PIAS) has developed a way of dramatically reducing the cost of these thermal systems (PIAS 2004).

The introduction of light emitting diodes (LED) has meant that many buildings are now replacing their traditional fluorescent light tubes (FLT) with much more efficient LEDs. In the past, the used FLT were simply discarded and sent to landfill. The Pacific Institute has developed a process for converting used FLTs into tube-type solar collectors. An overview of this process is shown in Figures 8-17 through 8-20.

1)   Step One: Discarded Fluorescent Tube

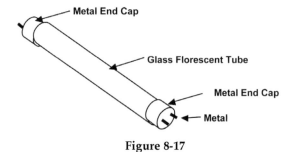

Figure 8-17

2)   Step Two: Fluorescent Tube With End Caps Removed
After the removal of the end caps, the tube is cleaned to remove the phosphor and mercury from the inside of the tube, leaving a clean, transparent glass tube.

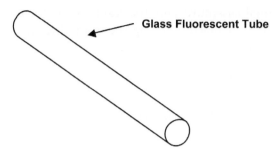

Figure 8-18

3) Step Three: Manufacture Solar Collection Plate

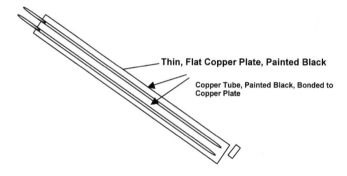

Figure 8-19

4) Step Four: Insert Solar Collection Plate Into Glass Tube

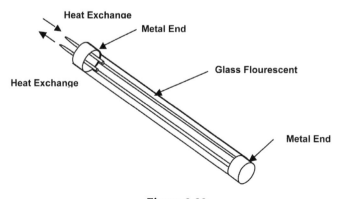

Figure 8-20

After the final assembly step, the result is an efficient, very low cost, glass tube solar collector, among the most efficient solar energy collection systems available.

Another approach to capturing solar energy is by the use of concentrating systems. While these systems are much more expensive than flat panels or tube collectors they can generate much higher temperatures in the working fluid. A typical large scale concentrating solar system is shown in Figure 8-21.[23]

Using absorption chilling technology, solar thermal collectors can be used to power a building air conditioning system. The warmed water from the panels is used to drive a thermodynamic process using a refrigerant such as lithium bromide that is non-toxic and not a green house gas. The process generates chilled water that is suitable for cooling small-scale residential and commercial buildings.[24]

A greatly simplified schematic of the absorption chilling process using lithium bromide as a working fluid is shown in Figure 8-22. The hot water is supplied from the solar collectors and the cooling water

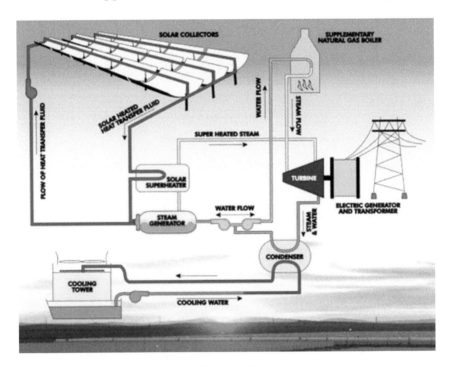

Figure 8-21

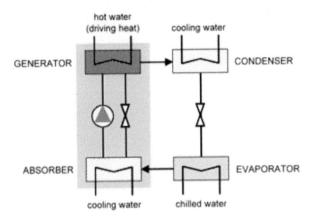

Figure 8-22

comes from the building's municipal water supply or from ground water. The chilled water is sent to "cooling panels" mounted on the ceiling of each room, effectively cooling the building.

The cooling effect is based on the evaporation of the refrigerant (water) in the evaporator at very low pressures. The vaporized refrigerant is absorbed in the absorber, thereby diluting the H2O/LiBr solution. As shown, to make the absorption process efficient, the process has to be cooled. The solution is continuously pumped into the generator, where the regeneration of the solution is achieved by applying driving heat (e.g. hot water from the solar panels). The refrigerant leaving the generator by this process condenses through the application of cooling water in the condenser and circulates by means of an expansion valve again into the evaporator.

WIND ENERGY

A wind turbine is a device for converting the kinetic energy contained within a moving stream of air into rotational motion. The rotational energy can then be converted into electrical energy. The power available is proportional to the square of the area swept out by the rotor. The application in major wind farms is well known, but a relatively recent development is the design of "building integrated" small wind turbines (BIWT), that in the near future could be integrated into major commercial buildings.

Figure 8-23 is a picture of a proposed BIWT[25]. An alternative approach is to mount relatively small wind turbines on a building's roof as shown in Figure 8-24.

Figure 8-23

Figure 8-24

Wind turbines are usually shipped as "turn-key" packaged systems that include the rotor, generator, turbine blades, and a drive or coupling device. Most systems are not direct drive, but have a gearbox and an induction (asynchronous) generator in a single unit behind the turbine blades.

Vertical axis wind turbines (VAWT) represent a design which has the potential to increase the application and efficacy of small-wind power for on-site generation. In small-wind, the VAWT can have a better energy density (kW/sq. ft of footprint) than the horizontal axis turbine, and may be especially advantageous for building sites. This is because the power production can be maximized per unit area and, due to the light weight of the equipment, can be more easily installed with existing roof structures. The aesthetic advantages of the VAWT cannot be underestimated as a factor as well. VAWTs can produce anywhere from 1kW-5kW per unit, a substantial increase over horizontal axis wind turbines (HAWT) with the same footprint and can often be grouped closer together. Figures 8-25 and 8-26 are two examples.

Figure 8-25[26]

Figure 8-26[27]

Almost all large commercial systems are horizontal axis and use a doubly-fed induction generator (DFIG) where the stator is directly coupled to the grid while the rotor is connected to the grid via an AC converter. However, one of the big disadvantages with wind energy is that it is a highly intermittent form of energy. If the wind speed falls below the minimum required to rotate the blades, electrical generation may come to a complete stop. Some VAWT designers have addressed this issue by engineering the turbines to rotate at very low wind speeds (5-10 mph) and still produce electricity.

As opposed to large turbines installed in remote locations, there are a number of disadvantages associated with BIWT. These include:

- There is a significant amount of turbulence around buildings that significantly reduces the efficiency and power output of the turbines.

- The maximum size of rotor that can be installed is necessarily limited.

- There are a number of associated engineering issues including vibration, ice accumulation in cold climates, noise etc.

- Grid-connected applications of large wind turbines typically range in cost from $0.04 to $0.10 per kWh. Small wind turbine energy costs range from $0.07 to $0.15 per kWh. For standalone applications of small wind turbines, energy costs typically range from $0.08 to $0.30 per kWh.[28]

• The systems are difficult to design properly because CFD modeling of urban wind movement is a real challenge. It is almost impossible to accurately predict because every environment has local characteristics that affect air flow.

Some of these disadvantages can be overcome by designing the building(s) and the integrated large-scale wind turbine at the same time in order to optimize their performance as a combined system. The problem is that the addition of a large capacity wind turbine adds significantly to the complexity and cost of the project.

As an example of some of the possible problems associated with this type of project, the vibrations caused by the rotors and gears need to be isolated from all occupied spaces. Another problem in northern climates is ice shedding. This is an unacceptable safety risk for turbines in urban areas where pedestrians are walking at the foot of tall buildings. While de-icing technologies do exist, they will significantly reduce the efficiency of the turbine.

Some designs have used ducted turbines mounted between two building towers. These systems will have increased efficiencies if the wind happens to be blowing from the optimum direction, but performance will fall to zero as the wind shifts to a direction perpendicular to the rotation axis of the blades. Also, placing turbines within a building structure will inevitably reduce floor efficiency and may introduce building code and safety issues.

It can be concluded that if it is desirable to incorporate sustainable onsite wind energy generation into a building, a thorough analysis of footprint available, local wind conditions and flow obstructions, kW output and regulatory requirements for installation must be done to compare efficacy versus PV-generated electricity.

BIOMASS ENERGY

Biomass is commonly defined to be the biodegradable fraction of products, waste and residues from agriculture (including vegetal and animal substances), forestry and related industries, as well as the biodegradable fraction of industrial and municipal waste. It cannot be considered as a real DPG technology, but simply as a fuel, and it can be seen as an alternative to the use of conventional fossil fuels.

In the combustion process, wood and other biomass materials are gasified to be used in electricity generation. Biomass can be combusted directly to generate the thermal energy to operate conventional steam turbines or it can be combined with another traditional fuel such as coal or oil to increase its energy content and its combustion efficiency. It can also be converted into liquid or gas and used as a fuel for any of the available DPG technologies including microturbines, gas turbines, fuel cells, ICEs, Stirling engines etc.

Systems using biomass as a solid, liquid and gaseous fuel are compatible with various CHP systems. The advantage is that various waste products from households, industry and agriculture can be used as fuels. Distributed biomass-fueled CHP systems can provide important environmental benefits including reduced greenhouse gas emissions and reduction in the consumption of fossil fuels. The combustion of biomass generally produces less toxic gases such as NOx.

The overall efficiencies of small biomass combustion systems are relatively low, so commercial plants tend to be quite large, with capacities of 20 MW or more. Because of this, biomass fueled systems may not be suitable for green building applications, and are better for large centralized power plants.

CONCLUSIONS

Based on the issues highlighted in this chapter, some technical recommendations for possible actions for DPG at the generation, transmission and distribution system level can be proposed:

• Cogeneration systems significantly increase the efficiency of a system as contrasted with systems that only produce process heat, or that only generate electrical power.

• Careful planning is necessary to outline the present and future evolution of the electricity and process heat systems or in order to avoid integration and distribution problems down the road.

- The DGP system may be subdivided into a number of subsystems by an islanding procedure. Each subsystem has to be eventually able to balance supply and demand effectively (i.e. be self-sufficient) for two reasons: First, to be able to decouple from the interconnected system and to continue running in case of large and widespread disruptions; and second, to reduce the impacts (in terms of control actions and disruptions) on any upstream systems.

- The provision of auxiliary services by DPG, especially by units interfaced via a power electronic converter or synchronous machines, is necessary for a complete integration of a DPG system into the building. Such a requirement is necessary both for technical reasons (DPG may take the place of a large generation facility in providing equivalent operation flexibility and reliability) and economic requirements (DPG investments may offer shorter payback periods by also marketing auxiliary services).

- The development of cost effective and efficient high-capacity energy storage systems, based on various technologies, may play a key role in facilitating a larger penetration of DPG systems.

- Apart from technical considerations, the broad introduction of DPG systems will depend on various regulatory and environmental issues. These aspects need to be carefully examined to properly manage the transition from strictly grid-connected buildings, to high-performance buildings which balance centrally supplied power and on-site distributed energy.

References
1. Also referred to as: Distributed Energy Generation (DEG)
2. Distributed Generation: a Definition, Ackermann, T., Andersson, G., Söder, L. Electric Power Systems Research, Vol. 57, 2001, pp. 195-204.
3. SEE: The Role of Energy Storage with Renewable Electricity Generation, NREL Technical Report, Denholm, P., Ela, E., Kirby, B., Milligan, M., 2010. http://www.nrel.gov/docs/fy10osti/47187.pdf, and: Energy Storage for Power Systems (2nd Edition), Ter-Gazarian, A., The Institution of Engineering & Technology, 2011. http://digital-library.theiet.org/content/books/po/pbpo063e
4. Assessment of Distributed Generation Technology Applications, Resource Dynamics Corporation, 2001. See: http://www.distributed-generation.com/Library/Maine.pdf

5. Westport Natural Gas Engines: http://www.westport.com/
6. SEE: http://www.slimfilms.com/
7. SEE: Electric Power Research Institute: http://www.epri.com/search/Pages/results.aspx?k=recuperated%20microturbine%20system
8. SEE: http://www1.eere.energy.gov/hydrogenandfuelcells/fuelcells/basics.html
9. SEE: DOE Fuel Cell Technologies Program Record http://www.hydrogen.energy.gov/pdfs/12020_fuel_cell_system_cost_2012.pdf
10. SEE: http://www.pacificinstitute.com/alteng/alteng1.html
11. SEE: http://en.wikipedia.org/wiki/File:Example_of_cylindrical_flywheel_rotor_assembly.png
12. SEE: The Stirling Engine Manual, Rizzo, G., Camden Miniature Steam Services, 1995.
13. SEE: http://powerelectronics.com/discrete-power-semis/renewable-energy-through-micro-inverters
14. One of the most advantageous approaches is a Power Purchasing Agreement (PPA) for the building owner. SEE: *How to Finance Energy Management Projects*, Woodruff, E., Thumann, A., CRC/The Fairmont Press, GA, 2013, ch. 5.
15. Solar Power Plant, Nellis Air Force Base, Nevada, 13.5 MW (AC) with Single-Axis Trackers. SEE: http://www.energybc.ca/profiles/solarpv.html, and : http://www.nellis.af.mil/shared/media/document/AFD-080117-043.pdf
16. SEE: http://www.pv-magazine.com/news/details/beitrag/pv-microinverters-and-power-optimizers-set-for-significant-growth_100008296/#axzz2SiDbzLlV
17. SEE: Power per unit area: http://www.onyxsolar.com/power-per-unit-area.html
18. CIS/CIGS: Copper-Indium-Selenium/Copper-Indium-Gallium-Selenium
19. SEE: http://www.nrel.gov/rredc/pvwatts/grid.html
20. A LBL report indicates that the median installed price of PV systems installed in 2011 was $6.10 per watt (W) for residential and small commercial systems smaller than 10 kilowatts (kW) in size and was $4.90/W for larger commercial systems of 100 kW or more in size. SEE: http://newscenter.lbl.gov/news-releases/2012/11/27/the-installed-price-of-solar-photovoltaic-systems-in-the-u-s-continues-to-decline-at-a-rapid-pace/
21. SEE: http://physicsworld.com/cws/article/news/2010/jun/24/quantum-dots-for-highly-efficient-solar-cells
22. Size-dependent band gap of colloidal quantum dots, Baskoutas, S., Terzis, A., F., Journal of Applied Physics 99: 013708, 2006. doi:10.1063/1.2158502
23. http://goodspeedupdate.com/date/2008/03
24. SEE: http://www.scusolar.org/technology.thermal
25. Bahrain World Trade Center: the turbines produce 1100-1300 MW of power annually, about 10-15% of the building's requirements. SEE: http://www.treehugger.com/renewable-energy/worlds-first-building-integrated-wind-turbines.html
26. 3.6KW VAWT with dual rotors. SEE: http://www.be-technologies.org/
27. SEE: http://www.technologyreview.com/news/513266/will-vertical-turbines-make-more-of-the-wind/
28. SEE: https://www1.eere.energy.gov/femp/technologies/renewable_wind.html

Chapter 9

Mechanical & Electrical Equipment

Tim Janos, C.E.M., AEE

INTRODUCTION

The buildings sector[1] in the US accounts for approximately 2.2% of petroleum consumption, 56% of natural gas consumption, and 74% of electricity consumption.[2] There are major opportunities to reduce on-site energy consumption at buildings, especially at peak load, by addressing waste reduction and efficiency improvements in mechanical and electrical (M&E) equipment. Cutting overall demand and peak load demand from buildings takes pressure off base-load power plants and reduces pressure on the grid as a whole.

As we have seen, close attention to cutting waste, performing audits, commissioning, building energy analysis, improving the efficiency of the envelope and possibly adding on DEG are essential to making buildings high-performance. And all of these measures should, ideally, be reviewed before assessing the design, retrofit, upgrade or replacement of mechanical and electrical equipment. A highly efficient envelope with good daylighting will significantly reduce the energy requirements for HVAC and electrical lighting, thereby reducing capital outlays, on-going O&M expenses and energy costs.

There are literally hundreds of HVAC and lighting systems available and in use, and obviously, given space requirements, we will have to be very selective about what we discuss here. Proceeding in accordance with our previous approach, we will address basic principles in the major system types which can be applied to high-performance improvements. Some very general principles which we can apply are as follows:

CLOSE THE LOOP   '

[1]   cut energy losses through distribution systems: pipes, ducts, wires, etc. (insulation, flow efficiency, leak detection)
[2]   increase motor and pump efficiencies
[3]   improve the luminous efficacy (lumens/watt) of electrical lighting
[4]   optimize the operation of M&E equipment with building management systems (BMS) and energy management systems (EMS)
[5]   recover and reuse waste heat wherever possible

The most important standards for adopting high-performance measures in M&E equipment are certainly ASHRAE Standard 90.1 and 189.1[3]. These standards, or sections of them, have been integrated into the model codes from the ICC including the recently released International Green Construction Code. So we will consider their guidelines for some of the controlling points in the discussion. Obviously, recommendations from these standards about electric lighting are derived from other sources as well, including the DOE and the Illuminating Engineering Society.[4]

HEATING, VENTILATING & AIR-CONDITIONING

We will confine our discussion primarily here to the most common types of HVAC: central heating and cooling systems with distribution trees of air, water, or both.

The vast majority of HVAC systems in a building, with numerous variations, are going to consist of:

[1]   an air handling and air or water/steam distribution system
[2]   a heating system employing a boiler or furnace
[3]   a cooling system employing a chiller or DX packaged unit (rooftop unit [RTU])

Codes and standards in commercial buildings also require a number of add-ons to these systems. In particular the codes require the installation of outdoor air economizers to provide sufficient ventilation air for the occupants and to maintain the correct air balance for building exhaust systems. High-performance building standards require heat recovery on the exhaust systems and further require air flow measuring devices on the outdoor air intake to be assured of minimum required ventilation standards. And, in almost all cases, to optimize these systems and make them high-performance, they should be controlled by a building automation system (BMS or EMS).

HEATING

Boilers

Most commercial office buildings with central fan-forced air distribution, are furnished with a perimeter heating system which accommodates heat loss through the envelope of the building. These systems vary from standard radiant fin-tube to hot water reheat coils in fan powered variable air volume boxes. Since these represent the majority of use, we will concentrate our discussion on the central heating plants used in most of these buildings.

Creating steam involves a change of state, and the energy required to change a pound of water into steam at 212°F is approximately 970 Btu. This reflects both the sensible heat and the latent heat due to the change of state. This makes the heat transport capabilities of steam very significant, although the trade-off comes in the difficulty of modulating and controlling the steam to keep tenants comfortable. Even today we find that steam systems are operated to supply a heat exchanger which then converts the steam energy into hot water and which is pumped to perimeter fin tubes. This is inherently wasteful and should be avoided if at all possible. Steam boilers are fairly common in older facilities sometimes coupled with perimeter steam radiation. It is still not unusual to see older buildings utilizing this heating technology as evidenced by windows opened in the dead of winter to counteract the tendency to overheat the exterior rooms. This is a commentary on waste in old steam boiler systems: overcapacity and oversupply of the steam for the demand required.

With existing, lower efficiency hot water boilers the only real options are either better controls—i.e. Indoor/outdoor reset controls or, perhaps, more efficient pumping systems. This, of course, assumes that the basic maintenance in terms of boiler tuning has been performed. Naturally, there are some subtle pumping modifications that also work. For example, many people use a throttling valve to balance the water flow in the system. This is the equivalent of a discharge damper on a fan as it has no horsepower reduction. Correct action is to open the throttling valve and install a VFD on the centrifugal pump.

High efficiency boilers have come a long way over the last decade, and the old standard modular boiler plant is no longer representative of efficient heating plant design. Condensing boilers[5] offer the highest possible turndown ratios[6], as much as 20:1, and therefore set the new standards for performance and heating efficiency with this type of heating system.

Figure 9-1

Older boilers have seasonal efficiencies below 75 percent and some new medium-efficiency boilers can achieve as much as 85 percent. However, condensing high-efficiency boilers, like the one shown in Figure 9-1[7], can achieve seasonal efficiencies as high as 96 percent. High efficiency boiler heating plants have become a relatively standard option for both new and retrofit installations, because of the life cycle cost savings. The key difference between high- and mid-efficiency boilers is their ability to condense water from combustion products in order to extract as much heat as possible. Condensing boilers are generally built with more expensive materials increasing their initial cost by approximately 15-18%. However, the energy efficiency gains over an older boiler will amortize the extra costs in 5-10 years.

In existing buildings, standard hydronic heating plants were designed to operate with a high supply temperature and a small temperature drop between the supply and return water. In comparison, high-efficiency boilers operate at peak efficiency when the return water temperatures are between 90°F and 140°F. Condensing boilers require a low return water temperature to operate at their highest efficiency and so these systems should be designed with lower flow rates. Heating coils and radiators should be sized for a higher rate of heat transfer at lower supply water temperatures.

Condensing boilers also function with smaller venting pipes, although more expensive stainless steel is required for larger boilers. Smaller systems can use PVC pipe, which can be directly vented to sidewalls.

If you are retrofitting a boiler your distribution and heat exchange systems must be thoroughly evaluated for compatibility. Some common characteristics of modern condensing boilers include stainless steel construction to withstand the low pH of the continuously condensing flue gases and a high turndown ratio for efficient operation at low loads.

What follows is a case study with an energy assessment[8] for a boiler replacement:

5.10.2 Item 2: Replace Boiler
ECO[9] Overview

The majority of the Community House is heated with a circulating hot water system with radiators in each room. Control on the radiators is with manual valves. Other than the office area which is used 40 hours a week, the rest of the areas conditioned by the boiler are only used about an hour a week. The boiler is controlled by a thermostat that turns the boiler on and off. This unit is at the end of its life and is scheduled to be replaced in 2012. No unoccupied setback is utilized.

A savings can be realized by replacing the 325,000 Btu boiler and upgrading the system to save more energy. Savings can be realized by making the following upgrades:

- Replacing shutoff valves on radiators with thermostatic control valves
- Utilize temperature setback during unoccupied hours.
- Replace old boiler with a high efficiency condensing boiler.

Assumptions for the retrofit are as follows :
- A total of 8 radiators are affected. Bypasses are not required to be installed around each radiator

- 72°F is the normal occupied space temperature. The spaces will be setback to 60°F in the winter.
- The boiler is sized to run full time during worst case loading conditions.
- The basement of the Community House is utilized 8 hours per day, 5 days per week. It is assumed that the normal occupied space temperature will be utilized 12 hours/day to accommodate workers who deviate from the standard workday and to allow time for morning warm-up/cool down. Holidays are included in the setback schedule. The banquet hall is rarely used; it is assumed that this area will be maintained at 60°F full time in the winter.
- Installation scope is as follows: Install a condensing type boiler with new PVC flue piping. An electronic thermostatic valve with sensor and thermostat will be installed on each radiator. A Honeywell Hometronic Manager (or similar product) will be used as a building automation system. This will allow each thermostatic radiator valve to have energy optimized temperature control.
- Because the boiler is already scheduled to be replaced, the cost shown for the boiler upgrade is only the cost difference between a typical boiler and a condensing boiler.

The evaluation calculation Table 9-1 identifies potential savings that would result from performing this ECO.

Commercial buildings very often require cooling in the "core areas" while, at the same time, during the heating season, they require heating at the perimeter of each floor. As discussed previously, hot water fin-tube or fan coil units are fairly common as these permit adjustment at the location of heat exchange which can significantly increase tenant comfort. Naturally then, since commercial buildings can experience seasonal situations where both heating and cooling are simultaneous, dynamic and efficient zoning control is required to avoid energy wasted in the mixing of these conditions.

Furnaces

High efficiency condensing technology is a vital part of high-performance furnaces resulting in an annual fuel utilization efficiency (AFUE)[10] of as much as 95%. The latest innovative furnaces incorporate modulating gas valves to match the space loads, along with the variable-speed ECM (electrically commutated motor—i.e. variable speed) blower motor and variable-speed inducer motor. Additionally, these furnaces provide very tight temperature control compared with old single stage furnaces.

Table 9-1

Natural Gas Cost/MMbtu	$11.26
Electric Cost/kwh	$0.130
Heating Efficiency (Existing)	70%
Heating Efficiency (Condensing Boiler)	93%
Occupied Temp °F	72
Winter Setback Temp°F	60

Unit CFM	Out Door Temp	Total Yearly Hours			Btu/yr New	Night Setback Reduction	Btu/yr New Net		MMBtu/yr Saved	$
0	97	15								$
0	92	121	0	0	0	0	0	0	0	$
0	87	241	0	0	0	0	0	0	0	$
0	82	324	0	0	0	0	0	0	0	$
0	77	647	0	0	0	0	0	0	0	$
0	72	779	0	0	0	0	0	0	0	$
0	67	986	0	0	0	0	0	0	0	$
0	62	738	0	0	0	0	0	0	0	$
0	57	452	0	0	0	0	0	0	0	$
0	52	611	52409	45745649	34432209	16579345	17852865	27892785	27.89	$ 314
0	47	584	104818.2	87448312	65821310	28524070	37297239	50151072	50.15	$ 565
0	42	727	157227.3	163291753	122907771	47344747	75563024	87728729	87.73	$ 988
0	37	619	209636.4	185378442	139532160	47029984	92502177	92876265	92.88	$ 1,046

(Continued)

Table 9-1 (Concluded)

0	32	710	262045.5	265788961	200056207	57797060	142259147	123529814	123.53	$ 1,391
0	27	481	314454.5	216075195	162637243	39155473	123481770	92593425	92.59	$ 1,042
0	22	351	366863.6	183955909	138461437	26668052	111793385	72162524	72.16	$ 812
0	17	206	419272.7	123385974	92871163	13415430	79455733	43930241	43.93	$ 495
0	12	96	471681.8	64687792	48689736	4688888	44000848	20686944	20.69	$ 233
0	7	39	524090.9	29199351	21978006	1058256	20919750	8279601	8.28	$ 93
0	2	33	576,500	27177857	20456452	0	20456452	6721406	6.72	$ 76
	Total	8760							627	
								Total Yearly Savings		$ 7054

Description	Qty	Material Cost	Labor Cost	Installed Cost (incl O&P)	Simple Payback (Years)	Return on Investment
Electronic Radiator Controller	8	$ 2,400	$ 400	$ 3,500		
Programmable T–Stat, wiring	1	$ 3,000	$ 1,000	$ 5,000		
Boiler Upgrade	1	$ 4,438	$ 1,738	$ 7,719		
Boiler Flue Piping (6" PVC) (ft)	100	$ 2,450	$ 1,815	$ 5,331		
		$ 12,288	$ 4,953	$ 21,550	3.1	33%

These furnaces operate very similarly to the high-efficiency condensing boilers discussed earlier in this section, when considering the gas burner, heat exchanger, and the additional energy savings resulting from varying the supply fan speed and the inclusion of the draft inducer. In general, the application of these ultra-high efficiency furnaces is limited by their heating capacity which usually does not exceed 200,000 Btu, and their air handling capabilities which are general designed for residential and light commercial retail strip store applications.

FEMP[11] provides the analysis shown in Table 9-2 of the net benefit that might be obtained by using the best available furnace technology for a specific application.

Table 9-2

Cost-Effectiveness Example(a)			
Performance(b)	Base Model	Required	Best Available
Annual Fuel Utilization Efficiency	80%	90%	95%
Annual Natural Gas Use	750 therms	660 therms	635 therms
Annual Natural Gas Cost	$750	$660	$635
Annual Natural Electricity Use	1,200 kWh	990 kWh	225 kWh
Annual Natural Electricity Cost	$96	$80	$18
Lifetime Energy Cost(c)	$11,730	$10,230	$9,030
Lifetime Energy Cost Savings	-	**$1,500**	**$2,700**

a) From GAMA's Consumer's Directory of Certified Efficiency Ratings for Heating and Water Heating Equipment.
b) More efficient products may have been introduced to the market since this Specification was published.
c) Lifetime energy cost is the sum of the discounted value of annual energy costs based on average usage and an assumed furnace life of 20 years. Future natural gas and electricity price trends and a discount rate of 3.0% are based on federal guidelines (effective from April, 2008 to March, 2009).

Cost-effectiveness Assumptions

Annual energy use in this example is based on the standard DOE test procedure for a non-weatherized furnace configured with an upward airflow and heating capacity of 72,000 Btuh. Operating hours are assumed to be 2,080 hours per year. The assumed price for natural gas is $1.00 per therm and electricity is $8¢ per kilowatt-hour, the average rates for federal facilities throughout the United States. The efficiency of the base model meets current US DOE appliance standards. The efficiency of the required model meets this Specification and is equipped with a standard fan motor.

The best available represents the most efficient product on the market for this size class that also meets the CEE/GAMA annual electricity use (EAE) criterion.

Using the Cost-Effectiveness Table

In the example shown above, the required furnace is cost-effective if its purchase price is no more than $1,500 above that of the base model. The best available is cost effective if its purchase price is no more than $2,700 above that of the base model.[12]

Electric Heating

In some climates, electric resistance heating may be used in lieu of gas-fired boilers and fintube. The advantage is clearly the installation cost difference with simple electric heating systems which could include baseboards, pedestal heaters, or possibly ceiling radiant panels. Occasionally, fan-powered VAV terminals with electric heating coils are used to heat perimeter offices. This provides individual zone control over the heating and generally is used where the engineer has determined that the heating needs of the perimeter offices are not significant.

Since electric resistance heat is more costly to operate than gas heat in most of the country, electric heat tends to be applied where the heating load is the lowest, i.e. the warmer climates, and the electric rates are most favorable. An additional important factor with electrical perimeter heating is that it is very controllable, and with the aid of a good building automation system, several energy conserving strategies, such as demand shedding or load rotation might be implemented if electrical demand charges are a significant part of the electrical tariff.

While not found as often, radiant electric heating panels[13] claim to provide improved comfort and substantial energy savings as indicated in the following link to a case study provided by the manufacturer. As all savings are tariff and application dependent, sufficient investigation and analysis is critical.

Heat Pumps

Although air-to-air electric heat pump systems are not generally known for being high-performance and can be problematic at low temperatures, they are sometimes a good choice for individual zoning systems (like hotel/motel applications) or in temperate climates where good envelope design maximizes natural ventilation and passive heating and cooling. They are most efficient in cooling applications, and for heating in moderate climate conditions.

An interesting approach to high-performance with heat pumps is the GAX gas-fired absorber heat pump system (or gas absorption heat pump (GAHP)). This is not exactly a mainstream technology and still has somewhat limited applications but is a high-performance system to be considered since it does not use electricity and utilizes a relatively clean and inexpensive fossil-fuel energy source: natural gas.[14] This addresses the problem of inefficient electric resistance heating in electric heat pumps when temperatures drop below 40°F. There a number of advantages with these systems since they can be combined with ground-source heat pump systems and renewable energy systems such as solar hot water heating. See Figure 9-2.

Figure 9-3 shows a gas-fired heat pump unit for heating or cooling from ROBUR.[15]

Geothermal (GeoExchange)

Geothermal, geoexchange, or ground source heat pump (GSHP) systems can be a very cost effective means of heating and/or cooling a building and also provide hot water.[16] A GSHP uses the ground as a source instead of air as a source in the air-to-air heat pump. Since at certain minimum depths, temperatures in the ground are relatively stable,[17] the temperature differential from surface air temperatures can be used to drive heating and cooling cycles in mechanical equipment in the building. The GSHP can reduce heating and cooling requirements by anywhere from 30-70% compared with conventional HVAC systems.[18]

The majority of a geothermal heat pump system is located underground.[19] See Figure 9-4. Looped pipes made of copper or polyethylene are buried in either horizontal or vertical loops. At the time of this writing, Geothermal heat pumps are considered to be a renewable energy technology and the federal government offers a tax credit in the amount of 30% of the total installed cost of the geothermal system. This tax benefit can be the driving factor for many building owners in making a decision to utilize the technology. The decision to employ the GSHP also very much depends on local area conditions such as soil engineering, available land area, etc.

The FEMP and DOE provide detailed information on ground source geothermal heat pumps, their efficiencies and applications.

Figure 9-2

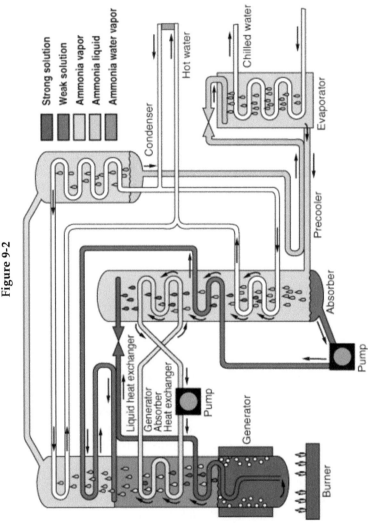

A flame from the natural gas burner heats a sealed pot containing a mixture of refrigerant and absorbent solution such as ammonia and water. The refrigerant is boiled out. Because the refrigerant—the ammonia—is in an enclosed chamber, heating also raises its pressure. The high-pressure ammonia vapor is then condensed, extracting heat from the refrigerant. The condensed refrigerant travels to the low-pressure evaporator, where the liquid refrigerant picks heat up from the environment—the cooling effect—and is turned once again into vapor, except now at low pressure and temperature. At the same time, the absorbent (water) from the generator, after the refrigerant is boiled out, travels to another heat exchanger called the absorber, which is at low pressure. The refrigerant vapor from the evaporator is next recombined with the water in the absorber. This recombining of the ammonia refrigerant and the water absorbent involves a chemical reaction that produces heat. This heat is removed from the absorber to increase GAX's thermal efficiency, and the now cool low-pressure mixture is pumped back to the generator, completing the process.

Figure 9-3

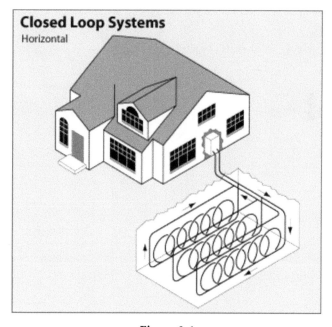

Figure 9-4

(*Continued*)

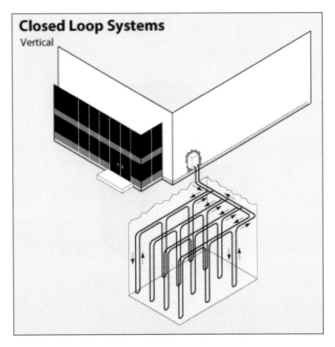

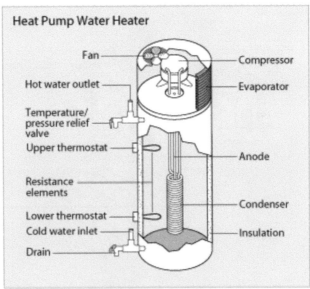

Figure 9-4 (*Concluded*)

COOLING

Rooftop Units (RTU)

Approximately half of all U.S. commercial space is cooled by self-contained, packaged rooftop air conditioning systems and almost all office buildings smaller than 200,000 ft. will fall into this category. These packaged units typically include the cooling components, the supply and return air fans, as well as the gas or electric heating sections. Commonly found on small office and low-rise buildings, retail centers, schools, hospitals, manufacturing facilities, the units today are much more flexible and efficient than even ten years ago. See Figure 9-5.

Single-zone low-pressure systems are still used for low-rise buildings, but most new commercial buildings are using variable air volume (VAV) to control loads to spaces with different heating and cooling needs. This permits the owner to receive all of the benefits of VAV reduced fan energy consumption while serving the facility's varied HVAC loads more efficiently than a single-zone system. The good news is that

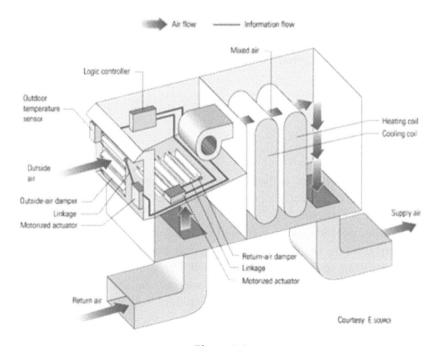

Figure 9-5

many Rooftop Units today can be equipped with full digital controls, integrated economizers, built-in demand controlled ventilation (DCV), multistage cooling systems, and VAV systems complete with factory installed variable frequency drives (VFD).[21]

Some manufacturers are offering larger rooftop units with integrated heat recovery wheels which can be very energy efficient in a college dormitory or similar application. Larger rooftop units typically approach a SEER in the range of 14. As a result of the U.S. Department of Energy's "RTU Challenge"[22] many manufacturers have produced rooftop HVAC units with integrated energy efficiency ratio (IEER)[23] higher than 18 as compared with older rooftop units at less than 10 SEER.

Figure 9-6 is the McQuay "Rebel"[24] cutaway illustrating the various energy savings aspects of this unit which is rated with a part load efficiency as high at 20.6 IEER, including Variable speed Daikin inverter compressor, variable speed Daikin heat pump, composite Daikin condenser fan(s), variable speed ECM fan motors, modulating hot gas reheat, MicroTech® III controls, stand-alone or hybrid heat options, electronic expansion valves, energy recovery wheel, and configurable as a 100% dedicated outdoor air, VAV, single-zone VAV, or CAV system.

While this is a significant improvement, the application for the high performance marketplace so far is primarily "big box" stores or possibly single-story small office buildings.

CHILLERS

Terminology
Ton

One ton of cooling is the amount of heat absorbed by one ton of ice melting in one day, which is equivalent to 12,000 Btus per hour, or 3.516 kilowatts (kW) (thermal).

Chiller performance is certified by the Air-Conditioning, Heating, and Refrigeration Institute (AHRI)[25], a manufacturers' trade organization, according to its Standard 550/590: *Performance Rating of Water-Chilling Packages Using the Vapor Compression Cycle*. Two efficiency metrics are commonly used for air-cooled chillers: full-load efficiency and part-load efficiency.

Full-load efficiency. Indicating the efficiency of the chiller at peak load, full-load efficiency is the energy-efficiency ratio (EER) measured

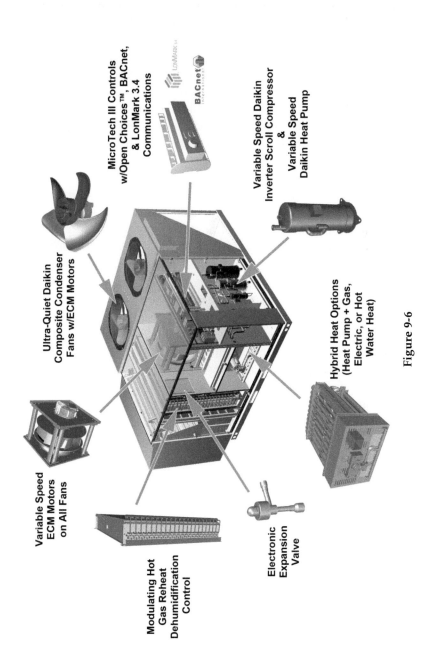

MicroTech III Controls
w/Open Choices™, BACnet,
& LonMark 3.4
Communications

Variable Speed Daikin
Inverter Scroll Compressor
&
Variable Speed
Daikin Heat Pump

Ultra-Quiet Daikin
Composite Condenser
Fans w/ECM Motors

Hybrid Heat Options
(Heat Pump + Gas,
Electric, or Hot
Water Heat)

Variable Speed
ECM Motors
on All Fans

Modulating Hot
Gas Reheat
Dehumidification
Control

Electronic
Expansion
Valve

Figure 9-6

at standard AHRI conditions. This is the ratio of the cooling capacity to the total power input, expressed in Btus per watt hour. A higher EER rating indicates higher efficiency.

Part-load efficiency. This metric indicates the efficiency of the chiller at part load, and is measured by integrated part-load value (IPLV)[26] of the particular AHRI part-load test conditions.

Both give the efficiency of the chiller using a weighted average formula referencing four operating load points (100 percent, 75 percent, 50 percent, and 25 percent) and are expressed in Btus per watt-hour.

Types

We'll begin by examining central station chiller equipment. Most newer office buildings larger than 200,000 square feet are going to be air-conditioned by means of a central chilled water plant. New chillers have typically been engineered to provide highly efficient operations under part load conditions. While chillers have typically been assessed in terms of coefficient of performance (COP), many use the standard of kW per ton performance as it is more readily understandable. Most office buildings have a particular load profile that includes the fact that maximum capability of the chiller is needed for less than 10% of the annual use. For this reason it is especially important to select chillers with very high part load efficiency. Chillers may either be air cooled or water cooled. A brief discussion of the characteristics of each follows. A schematic for an absorption chiller which is suitable for industrial or district applications is shown in Figure 9-7.[27]

Air-Cooled Chillers

Air-cooled chillers are most often found in facilities requiring less than 200 or 300 tons of chilled water capacity. They are very often a better choice than rooftop HVAC units as the chillers have a much greater "turndown" capability. That is, they are better able to match part-load cooling conditions than RTUs. These two factors enable air-cooled chillers to provide better humidity and temperature control than RTUs, which translates into greater occupant comfort and better efficiency.

Water-cooled Chillers

The use of the cooling tower as a heat rejection device gives water-cooled systems an efficiency edge over air-cooled systems. All chillers circulate chilled water to air-handler units where fans push air across

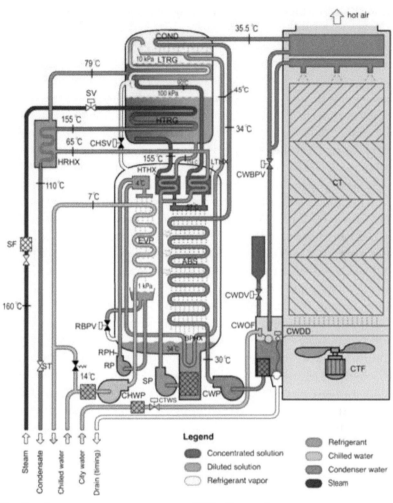

ABS = Absorber
BPHX = By-pass heat exchanger
CHSV = Cooling/heating switch valve
CHWP = Chilled water pump
COND = Condenser
CT = Cooling tower
CTF = Cooling tower fan
CTWS = City water switch
CWBPV = Cooling water by-pass valve
CWDD = Cooling water drain device
CWDV = Cooling water detergent valve
CWOF = Cooling water overflow
CWP = Cooling-water pump

EVP = Evaporator
HRHX = Heat recovery heat exchanger
HTHX = High-temperature heat exchanger
HTRG = High-temperature regenerator
LTHX = Low-temperature heat exchanger
LTRG = Low-temperature regenerator
RBPV = Refrigerant by-pass solenoid valve
RP = Refrigerant pump
RPH = Refrigerant pump heater
SF = Steam filter
SP = Solution pump
ST = Steam trap
SV = Steam valve

Figure 9-7

heat exchanger coils to deliver cooling. Because they circulate water, which is more energy dense than air, water-cooled chillers can offer a more efficient and effective cooling option than RTUs. Water-cooled chillers are more commonly used in buildings larger than 200,000 square feet, where the cooling load is large enough that increased efficiency gains offset the higher equipment cost of the cooling tower.

When selecting a chiller, spending more for greater efficiency can really pay off as in some areas of the country annual energy costs could be up to half of a chiller's purchase price. The reason part load variable efficiency may be the most important determining factor is simply that the full load efficiency among various models shows very little difference. The most significant difference is in the part load performance as measured by the IPLV. Chillers that use variable-frequency drives (VFDs) tend to increase the IPLV values even further and can produce chiller energy savings of 15 to 25 percent compared to standard models. Using VFDs at a chiller plant can produce large savings where the cooling load varies significantly.

In general, for conditions that require less than 300 tons of cooling capacity it is common to use chillers with screw or scroll compressors. While screw chillers dominate the upper end of this capacity range an important and relatively new technology shown in Figure 9-8 is magnetic-bearing compressors.[28] Magnetic bearing compressors which employ variable frequency operation are very efficient, power savings averaging 49% in case studies conducted by the Navy. Additionally since oil is not being circulated along with the refrigerant, an additional efficiency is obtained due to lack of de-rating of the refrigerant heat transfer capabilities by the oil. Applications needing more than 300 tons of cooling capacity are typically served by chillers with centrifugal compressors. Engineering an efficient chiller system is a challenging process in which many parameters affecting system efficiency and performance need to be considered, including the efficiency of auxiliary components such as pumps and fans.

ASHRAE 90.1-2007 Addendum m

This addendum reflects continuing improvements in VFD technology, which has improved chiller part-load efficiencies and encourages the use of higher-efficiency equipment. This publication ushered in several changes to the chiller requirements in ASHRAE Standard 90.1, the *Energy Standard for Buildings*.

Figure 9-8
Image courtesy of the Navy Techval Program

The biggest change was to replace a single compliance path with two different paths. Path A affects applications that spend a significant amount of time at full load. Path B affects applications that spend a signification amount of time at part load. This specification encourages the use of chillers with better IPLVs in part-load applications and full-load efficiencies in full-load applications. For either path, minimum requirements for both full load and IPLV must still be met.

Variable primary flow chilled water systems are an additional opportunity that should be considered particularly in any application where significant pumping energy is required. See Figure 9-9.

Variable-Primary-flow Systems

One big advantage of variable primary flow chilled water systems is the elimination of the constant flow primary loop which was normally decoupled from the secondary distribution group. This permits the use of two-way chilled water control valves on the air handlers ultimately resulting in reduced pumping horsepower due to the minimization of bypassing. These systems are especially effective in situations where there are significant amounts of part load operation.

In one case study, after installation of a variable primary flow chilled water system at the University of Arkansas at Fayetteville, results were documented as follows:

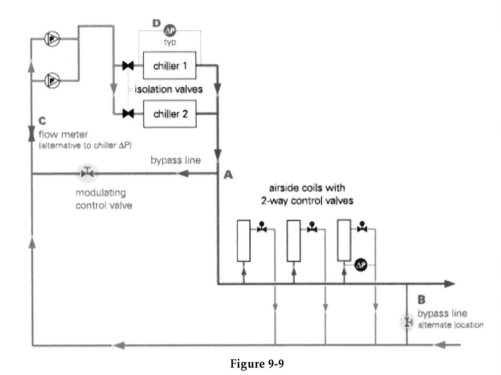

Figure 9-9

Annual energy savings of 1.9 million kWh and $67,000 in pumping power, and
$1,200,000 in annual chiller electrical and gas energy savings

AIR HANDLING UNITS (AHU) & DISTRIBUTION SYSTEMS

Air distribution systems for office buildings have become increasingly sophisticated and complex. Obviously the air distribution system must deliver conditioned air throughout the building in the most efficient manner possible. For most new commercial buildings, that means a variable air volume system (VAV) will be employed for the purpose of supplying individual zone temperature control as well as obtaining the benefits from fan speed control.

Any discussion about air distribution systems involves the following typical components: the air handling unit, the outdoor economizer as required by code, the air flow/fan speed control system for the air handler, ventilation control, minimum outdoor air requirement for ventilation, and exhaust air energy recovery.

The central station air handler is normally located in a mechanical room and includes chilled water coils for air conditioning purposes, often a separate hot water coil for heating purposes or perhaps a bank of electric heaters in certain climates, a supply fan blower section with motor, a return/exhaust fans section with motor, outdoor air and economizer dampers, and a mixing box with associated return/purge damper controls.

Current code requirements also include the necessity of adding air flow monitoring stations to assure that minimum outdoor air requirements are met. In some cases heat recovery may be required on the exhaust air. Since air handlers are either modular in nature, or completely custom-built, the manufacturers have become adept at providing these options as required. We would expect all new high-efficiency air handlers to be equipped with double wall insulated construction, high-performance low leakage damper sets, premium efficiency motors, and factory installed variable frequency drives. Depending on the manufacturer these air handlers are likely pre-integrated with digital controls.

The ability of a high-performance air handler to deliver energy savings is dependent upon the proper operation and interrelationship of all the components. Of particular concern should be the condition and adjustment of the outdoor air dampers. Outdoor air dampers that are not properly adjusted and controlled can provide a significant parasitic load on the system. Fortunately, current practices will result in the implementation of DCV or demand controlled ventilation as a practical matter in all high performance buildings. Demand controlled ventilation is permitted by ASHRAE Standard 62.1[30] and is supported by Table 9-3.

When a commercial building is designed, the engineers must consider the ventilation codes' maximum occupancy as well as the size of the space. Of particular note is the requirement for default occupancy assuming five people per every thousand square feet. Your outdoor air ventilation requirements are therefore driven by the maximum projected occupancy of the building. In the real world office buildings are dynamic places with employees constantly moving in and out of the space which creates an environment of variable occupancy and consequently variable ventilation loads.

Since outdoor air is expensive to heat, cool, humidify or dehumidify it is incumbent upon any good HVAC system to "right size" the

Table 9-3

Example: "Minimum Ventilation Rates in Breathing Zone", Source: ASHRAE 62.1-2007 and 62.1-2010, Table 6-1

Occupancy Category	People Outdoor Air Rate R_p (cfm/person)	Area Outdoor Air Rate R_a (cfm/ft^2)	Default Occupant Density (#/1000 ft^2)	Default Combined Outdoor Air Rate (cfm/person)
Educational Facilities				
Classrooms (age 5-8)	10	0.12	25	15
Classrooms (age 9 plus)	10	0.12	35	13
Science laboratories	10	0.18	25	17
Lecture classroom	7.5	0.06	65	8
Computer lab	10	0.12	25	15
Hotels, Motels, Resorts, Dormitories				
Bedrooms, living room	5	0.06	10	11
Lobbies/prefunction	7.5	0.06	30	10
Multipurpose assembly	5	0.06	120	6
Office Buildings				
Office space	**5**	**0.06**	**5**	**17**
Reception areas	5	0.06	30	7
Telephone/data entry	5	0.06	60	6
Main entrance lobbies	5	0.06	10	11

This table is extracted from ASHRAE Standard 62.1 covering ventilation rates. This table specifies minimum default outdoor air rates based on both the number of people occupying the space and the size of the space. For example if you looked at the highlighted listing for office space you would discover that the minimum outdoor air freight in terms of cfm/person is 17 cfm.

amount of outdoor air brought into the facility that needs to be treated. Standard 62 permits us to do this by using CO_2 as an indicator of the occupant load which provides us with the basis for demand controlled ventilation applications. ASHRAE Standard 62 actually provides three different means of complying with the standard. One method is to simply follow the tables and calculate the outdoor air based on those conditions and is sometimes referred to as the prescriptive method or the ventilation rate procedure. The second method is referred to as the indoor air quality procedure which is a performance based design that allows us to use indoor air quality sensors (CO_2) to maintain indoor air quality while adjusting the amount of outdoor air to measured occupancy condition.

A third provision, known as a natural ventilation provision allows for the use of windows or other means of delivering ventilation, although it does require a mechanical backup system should the other means be unavailable. It is unlikely that the natural ventilation provision will be used in most office buildings, unless it is subject to automation, since designers and engineers usually resist the idea of occupants manually controlling fresh air ventilation for obvious reasons.

The application of demand controlled ventilation requires knowledge of the regional air quality in accordance with the national ambient air quality standards, as well as a knowledge of local outdoor air conditions. Human occupants produce bio effluents including CO_2, water vapor, particulates, biological aerosols and VOCs (volatile organic compounds). Comfort (odor) criteria with respect to human bio effluents are likely to be satisfied if the ventilation results in indoor CO_2 concentrations less than 700 ppm above the outdoor air concentration. As a practical matter the outdoor CO_2 concentration is almost always less than 300 ppm.

It's important to note and understand that in order to comply with codes, the design engineer is required to design a system capable of bringing in all of the ventilation air needed to satisfy Standard 62 Prescriptive Maximums. To comply, ventilation systems often operate at a fixed rate based on an assumed occupancy (e.g., for an office, 17 cfm per person multiplied by the maximum design occupancy). Because of this, it is incumbent upon the operator of a high-performance building to be absolutely sure that the demand controlled ventilation system is well-maintained and properly adjusted to avoid the introduction of unneeded outdoor air. That air must be conditioned, resulting in higher energy consumption and costs than is necessary with appropriate ventilation. In humid climates, excess ventilation also can result in uncomfortable humidity, possibly mold or mildew growth, reducing indoor air quality (IAQ).

The other end of the spectrum is that a lack of adequate fresh air can make building occupants drowsy and uncomfortable. There are code specific requirements such as the number of sensors, a minimum of one sensor or sensing probe for each 10,000 ft² of floor space, as well as a requirement for dynamically measuring CO_2 by using this type of sensor located in the outdoor air intake to validate outdoor air conditions.

According to the Federal Energy Management Program (FEMP)[31], potential energy savings obtainable through a demand controlled ventilation system range from a low of $.10 per square foot to as high as $1.00 per square foot depending on the application. A secondary benefit will accrue to the high-performance building using DCV as an energy reduction method since the code has a mandate for outdoor air economizers. ASHRAE standard 90.1 mandates the use of outdoor air for "free cooling" when the outdoor air temperature is sufficiently below that of the indoor air temperature and the indoor air temperature is above the desired space setpoint. Current code requirements dictate that every air-conditioning system having a capacity equal to or greater than 33,000 Btus per hour must have an outdoor air economizer dependent on the climate zones. In some designs the engineer may have separate systems for the perimeter and for the core areas to achieve energy efficiency and optimum control of the space temperatures. One other type of application involves the use of dedicated outdoor air systems (DOAS)[32] in which a system is specifically engineered and applied to supply only the needed ventilation air of the spaces.

Underfloor Air Distribution

Another system a facility may employ is underfloor air distribution as shown in Figure 9-10. This is a newer technology where air is discharged upward resulting in more effective ventilation in the occupied zone and some stratification in the zones above occupancy. While the original versions of this were constant volume systems with adjustable air outlets in the floor, new versions in fact are able to employ variable air volume techniques to provide additional savings with reduced fan horsepower.

There are several advantages to this type of a system in a commercial building, including energy savings and better thermal comfort. Generally, the air is supplied at a temperature of 63°F instead of the typical 55°F air used in overhead VAV air distribution, resulting in energy savings at the supply side. The net results are also usually fewer complaints from occupants about space temperature due to the virtual minimization of stratification in the occupied zone. Additional benefits accrue from a remodeling standpoint as it is very simple and inexpensive to accommodate tenant changes since all the electrical wiring, communication wiring, and HVAC is underfloor.[33]

ASHRAE Standard 62 is under constant monitoring in their Standards Process Committee, and revisions as needed in the newest version will address required ventilation rate in occupied zones by incorporat-

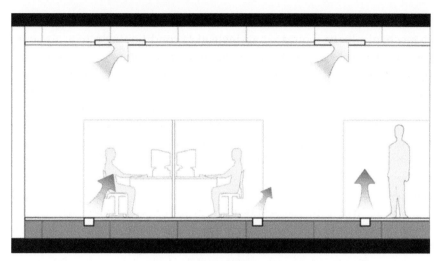

Figure 9-10

ing standards for those systems that utilize underfloor air distribution. Research has shown that the effectiveness of underfloor air distribution can result in less outdoor air required to adequately ventilate the occupied space.

OTHER EQUIPMENT MODIFICATIONS

Economizers

There are many engineers who believe that there are essentially two types of outdoor air dampers: Type I is the outdoor air damper that is already broken; Type II is the outdoor air damper that is about to break. One might think this is merely a humorous take on the situation but the reality is many surveys find that outdoor air economizers are routinely neglected from a maintenance standpoint. So they are frequently found to be substantially out of adjustment, either allowing excess outdoor air into the space to be conditioned, or they are effectively closed, contributing to the expected health and environmental issues associated with poor ventilation.

Exacerbating the situation is the fact that for a period of time there was an industry inclination towards using differential enthalpy controlled economizers. The principle behind differential enthalpy controlled economizers is that one can calculate when it is less expensive to mechanically cool outdoor air instead of conditioning return air coming back from the space. When these controls are brand-new there is an opportunity to obtain a measure of savings using differential enthalpy controllers. The problem with these controls is the fragility of the humidity sensing components. Even in the best case scenario, humidity sensors should be replaced every two years. Having an out of calibration humidity sensor used to determine differential enthalpy and then deciding whether to condition the outdoor air or return air is a risky proposition considering the limited life-span and effectiveness of the devices.

In an article in the ASHRAE Journal of November 2010 titled, "Why Enthalpy Controlled Economizers Do Not Work,"[34] the authors Stephen Taylor and C. Chang suggest quite strongly that more benefits are obtained with the implementation of dry bulb changeover[35] than the benefits that could be obtained from differential enthalpy controls. See Figure 9-11.

Packaged Unit (DX cooling) Outside Air Economizer

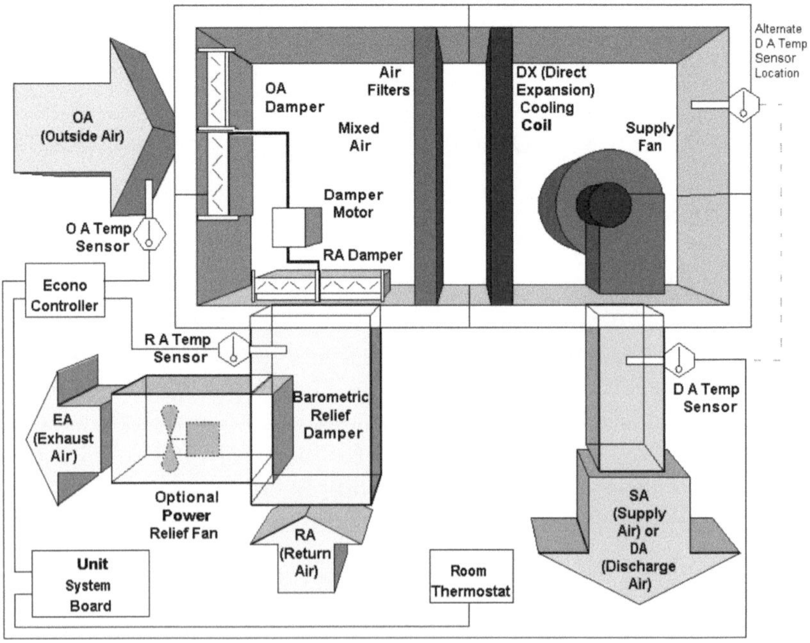

Figure 9-11

Another frequently overlooked issue associated with economizers is the fact that the outdoor air intake dampers are located either very close to the central station air or in the case of a packaged rooftop unit on the unit itself. When economizers operate to provide "free cooling" by opening the outdoor air dampers, virtually all of the duct resistance of the return air stream is eliminated causing the supply fan to perform in a different manner than it would if the suction side of the fan was dealing with all of the return air ductwork load. Complicating the situation, especially in the case of small package rooftop units, is the fact that barometric relief may be used to maintain building air balance.

Studies indicate that economizers which utilize barometric relief are less effective because the building doesn't adequately ventilate. Consequently, an effective outdoor air economizer system will be coupled with a combination return exhaust fan in order to maintain effective building air balance. It has been determined that the use of power exhaust to maintain building air balance greatly increases the effectiveness of the outdoor air economizer system in terms of effectively cooling the occupied space.

ELECTRICAL

Motors & Pumps

The Energy Independence and Security Act of 2007(EISA) went into effect on December 19, 2010 for new motors. This act raised the minimum efficiency level for 1–200 horsepower (HP) motors covered by EPACT-92 to the NEMA Premium® level. Motor Guide 1 gives us the required efficiencies in Table 9-4.

There are a couple of easy observations one can make by examining the table. The first is to notice that the bigger the horsepower, the more efficient the motor must be. If you examine column 1 you'll see that a 1 HP motor has a required efficiency of 77% whereas a 200 HP motor has a required efficiency of 95%. The second observation is that totally enclosed fan cooled motors (TEFC) are slightly more efficient than opened drip proof motors (ODP). Premium efficiency motors are important for a number of reasons, not least of which are the substantial energy savings achievable.[36] The Advanced Manufacturing Office of the DOE, in association with Washington State University, publishes a free software program: Motormaster+[37] which evaluates the usage profiles and performance characteristics of motors.

Table 9-4

NEMA MG-1 Table 12-12 Full-Load Efficiencies for 60Hz NEMA Premium® Efficient Electric Motors Rated 600 Volts or less (Random Wound)

Motor Horsepower	Nominal Full-Load Efficiency					
	Open Motors			Enclosed Motors		
	2 Pole	4 Pole	6 Pole	2 Pole	4 Pole	6 Pole
1	77.0	85.5	82.5	77.0	85.5	82.5
1.5	84.0	86.5	86.5	84.0	86.5	87.5
2	85.5	86.5	87.5	85.5	86.5	88.5
3	85.5	89.5	88.5	86.5	89.5	89.5
5	86.5	89.5	89.5	88.5	89.5	89.5
7.5	88.5	91.0	90.2	89.5	91.7	91.0
10	89.5	91.7	91.7	90.2	91.7	91.0
15	90.2	93.0	91.7	91.0	92.4	91.7
20	91.0	93.0	92.4	91.0	93.0	91.7
25	91.7	93.6	93.0	91.7	93.6	93.0
30	91.7	94.1	93.6	91.7	93.6	93.0
40	92.4	94.1	94.1	92.4	94.1	94.1
50	93.0	94.5	94.1	93.0	94.5	94.1
60	93.6	95.0	94.5	93.6	95.0	94.5
75	93.6	95.0	94.5	93.6	95.4	94.5
100	93.6	95.4	95.0	94.1	95.4	95.0
125	94.1	95.4	95.0	95.0	95.4	95.0
150	94.1	95.8	95.4	95.0	95.8	95.8
200	95.0	95.8	95.4	95.4	96.2	95.8
250	95.0	95.8	95.4	95.8	96.2	95.8
300	95.4	95.8	95.4	95.8	96.2	95.8
350	95.4	95.8	95.4	95.8	96.2	95.8
400	95.8	95.8	95.8	95.8	96.2	95.8
450	95.8	96.2	96.2	95.8	96.2	95.8
500	95.8	96.2	96.2	95.8	96.2	95.8

Motor Operating Cost Comparison

Motors, like automobiles, have performance ratings, depending on their size and configuration. Motor performance is measured by efficiency, automobile performance, by MPG.

	Automobile	*60 HP Motor*
• **Purchase Price**	**$25,000**	**$2,600**
• Annual Usage	12,000 miles	8,760 hours
• Efficiency	20 mpg	93.6%
• Energy Cost	$4.00/gallon	10¢ kWh
• **Annual Operating Cost**	**$2,400/year**	**$41,890/year**

The idea is to illustrate the fact that the actual purchase price of the motor is greatly overshadowed by the annual operating cost. We might pay $25,000 for an automobile which gets 20 miles to the gallon and which would cost us about $2400 a year in gasoline. A 60-horse motor might cost only $2600, be 93.6% efficient, but actually results in an operating cost of $41,890 per year. See Table 9-5

Motors are significant users of electricity in the US and globally. In fact it is estimated that in the USA more than 50% of all electricity is used by motors[38] and 47% of all electricity globally.[39] A premium efficiency motor is a function of improved design in the rotor, the stators, the conductor bars, better bearings, and better manufacturing.

Nearly every existing commercial building has numerous motors in air handlers, pumping systems, etc., which were manufactured prior to the new standards. Industrial buildings have even greater capacities. Since most motors operate a very high percentage of the time, the opportunity to save energy is substantial when considering motor replacement.

Belt Drives

Good V-belts typically achieve efficiencies in the 90 to 95 percent range. A worn belt, however, will considerably reduce efficiency due to slippage caused by slackening and the normal wear of the belt grip surfaces. Cogged V-belts are similar to standard V-belts, except that the normally flat underside has longitudinal grooves in it, allowing better grip and less slip than standard V-belts. They typically offer a 2 to 5 percent efficiency bonus.[40]

Less common are synchronous belts which combine toothed belts with grooved sprockets. The belts rotate in exact synchrony, eliminating losses from slippage and significantly reducing maintenance because the belts are non-stretch. These belts transmit power by engaging teeth which allows them to operate much more efficiently than V-belts, achieving efficiencies in the range of 97 percent to 99 percent.

A potential downside to both premium efficiency motors as well as synchronous belts is that they will actually increase the speed of the fan, resulting in increased airflow. Since fans are subject to the fan laws we must remember that it will also require more power from the motor. For example, increasing fan speed as a result of reduced belt slip or installing a Premium Efficiency Motor will result in a corresponding increase in RPM of the fan wheel and an increase in the volume of air that is de-

Table 9-5

Estimated Annual Energy Savings with NEMA Premium® Motors[1]

TEFC · 1800 RPM · Full-load Operation · Nominal Efficiency[3]

Estimated Annual Energy Cost = (Hp x annual operating hours x cost of electricity x 0.746) / (efficiency)

Motor Decisions Matter — www.motorsmatter.org

Learn more about saving money through sound motor management. For additional tools and resources, visit www.motorsmatter.org.

Enter Appropriate Values for:

Annual Operating Hours:	8000 hrs per year
Cost of Electricity:	10 ¢ per kWh

Motor Size[2] (hp)	Estimated Annual Operating Costs		Estimated Annual Energy Savings	Estimated Annual Operating Costs		Estimated Annual Energy Savings	Estimated Annual Operating Costs		Estimated Annual Energy Savings
	Pre-EPAct Motors	EPAct Motors	Pre-EPAct → EPAct	EPAct Motors	NEMA Prem Motors	EPAct → NEMA Prem	Pre-EPAct Motors	NEMA Prem Motors	Pre-EPAct → NEMA Prem
1	$778	$723	$55	$723	$698	$25	$778	$698	$80
1.5	$1,132	$1,066	$66	$1,066	$1,035	$31	$1,132	$1,035	$97
2	$1,477	$1,421	$56	$1,421	$1,380	$41	$1,477	$1,380	$97
3	$2,200	$2,046	$153	$2,046	$2,000	$46	$2,200	$2,000	$199
5	$3,582	$3,410	$172	$3,410	$3,334	$76	$3,582	$3,334	$248
7.5	$5,235	$5,001	$234	$5,001	$4,881	$120	$5,235	$4,881	$354
10	$6,964	$6,668	$296	$6,668	$6,508	$160	$6,964	$6,508	$456
15	$10,337	$9,837	$500	$9,837	$9,688	$149	$10,337	$9,688	$649
20	$13,487	$13,116	$371	$13,116	$12,834	$282	$13,487	$12,834	$653
25	$16,708	$16,147	$561	$16,147	$15,940	$207	$16,708	$15,940	$768
30	$19,982	$19,377	$606	$19,377	$19,128	$248	$19,982	$19,128	$854
40	$26,466	$25,669	$797	$25,669	$25,369	$300	$26,466	$25,369	$1,097
50	$32,683	$32,086	$597	$32,086	$31,577	$509	$32,683	$31,577	$1,107
60	$39,007	$38,256	$750	$38,256	$37,693	$564	$39,007	$37,693	$1,314
75	$48,811	$47,566	$1,245	$47,566	$46,918	$648	$48,811	$46,918	$1,893
100	$64,659	$63,153	$1,505	$63,153	$62,558	$596	$64,659	$62,558	$2,101
125	$80,911	$78,942	$1,969	$78,942	$78,197	$745	$80,911	$78,197	$2,714
150	$96,258	$94,232	$2,026	$94,232	$93,445	$787	$96,258	$93,445	$2,813
200	$127,658	$125,642	$2,016	$125,642	$124,075	$1,567	$127,658	$124,075	$3,583

Size[2] (hp)	Pre-EPAct Motors	NEMA Energy Efficient Motors	Energy Efficient → EPAct	NEMA Energy Efficient Motors	NEMA Prem Motors	Energy Efficient → NEMA Prem	Pre-EPAct Motors	NEMA Prem Motors	Energy Efficient → NEMA Prem
250	$158,386	$157,053	$1,334	$157,053	$155,094	$1,959	$158,386	$155,094	$3,293
300	$189,661	$187,673	$1,988	$187,673	$186,112	$1,561	$189,661	$186,112	$3,549
350	$220,803	$218,952	$1,852	$218,952	$217,131	$1,821	$220,803	$217,131	$3,672
400	$251,814	$250,231	$1,584	$250,231	$248,150	$2,081	$251,814	$248,150	$3,665
450	$282,993	$281,509	$1,483	$281,509	$279,168	$2,341	$282,993	$279,168	$3,824
500	$314,436	$311,482	$2,954	$311,482	$310,187	$1,295	$314,436	$310,187	$4,249

1. This chart provides an estimated comparison of annual energy costs for Pre-EPAct, EPAct and NEMA Prem motors. Actual costs and savings may differ from the values shown.

2. The break in Motor Size between 200 and 250 hp occurs because EPAct applies to motors up to 200 hp. Above that value, NEMA's Energy Efficient Motor specification has been used as the reference.

3. The nominal efficiency values used in these calculations are defined as follows: Pre-EPAct Motors: DOE's MotorMaster+ software version 4.00.01 (9/26/2003) "Average Standard Efficiency" motor defaults ; EPAct Motors: Energy Policy Act of 1992 ; Energy Efficient Motors: NEMA MG 1-2003 Table 12-11 ; NEMA Premium Motors: NEMA MG 1-2003 Table 12-12. A table of all efficiency values is provided as the second tab of this Excel Workbook. Estimated Annual Energy Savings Chart, available at www.motorsmatter.org.   11/1/05

livered. Recalling the cubic relationship between airflow and horsepower, a 3% increase in fan speed means that the horsepower required for the fan will increase by (1.03)3, or 1.093—a 9.3 percent increase. Energy-efficient synchronous belts can easily be incorporated into a standard maintenance program, and the savings generated greatly outweigh the slight increase in cost per belt.

Variable Frequency Drives and the Centrifugal Fan Laws

If the building is equipped with a variable air volume HVAC system then there is some method of varying the air flow output of the air handlers. If we ignore hydraulic or magnetic clutching assemblies, there are essentially three methods for air flow volume control.

These are the discharge damper, inlet vane damper, or variable frequency drive.

Discharge dampers are placed at the outlet of the blower assembly and while these dampers will result in reducing airflow they are completely ineffective in terms of reducing energy consumption for that blower and motor combination. In Figure 9-12 discharge dampers would typically be located at #1.

Discharge Dampers Inlet Vanes

In older built-up central station air handlers inlet vanes would be found on the suction side of the fan as illustrated by #2 in Figure 9-12. These are decidedly better than outlet dampers in terms of performance, provide decent airflow control and are very common. It would not be at all unusual in upgrading a facility to high performance to discover that the air handlers have this method of airflow control. Inlet vanes do provide some horsepower energy savings.

In #3 of Figure 9-13, VFD is applied to the electrical feed to the motor itself and manipulates the speed of the motor by modifying the frequency of the electrical supply. It is both effective and efficient. VFDs have become so cost competitive, that it is frequently recommended to replace a failed motor starter with a VFD. Even if it is determined that the motor needs to run near full load, any minor speed reduction that can be obtained yields big benefits. For example, a motor running at 80% of full speed uses about 50% of the horsepower due to the fan laws as shown in Figure 9-13.

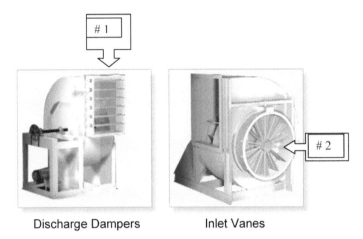

Discharge Dampers   Inlet Vanes

Figure 9-12

Figure 9-13

Variable Frequency Drives (VFD)

VFDs, sometimes referred to as variable speed drives, are electronic devices that influence the operation of the electric motor. Essentially variable speed drive will vary the frequency of the electricity supplied to the electric motor as a means of controlling the speed. Variable speed drives are carefully engineered to follow the centrifugal fan law demonstrated in Figure 9-14. This is the most cost-effective and efficient means of fan speed and airflow control.

The VFD default loading profile in Figure 9-15 shows clearly why variable frequency drives can be used effectively as a means of saving energy. The bell shaped curve demonstrates that the bulk of the air flow needed is in the 50-60% range.

The centrifugal fan laws follow.

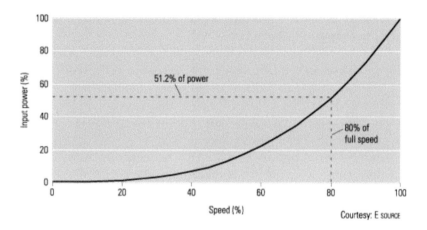

Figure 9-14

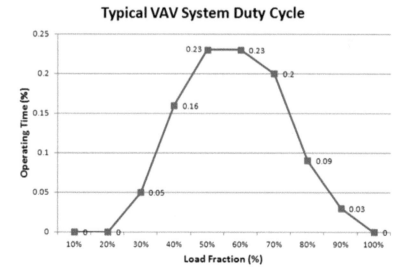

Figure 9-15

Electric Motor Management Fan Laws
(Centrifugal Devices ONLY)
- $CFM_2 = CFM_1(RPM_2/RPM_1)$ 1st law
- $SP_2 = SP_1 (RPM_2/RPM_1)^2$ 2nd law
- $HP_2 = HP_1 (RPM_2/RPM_1)^3$ 3rd law

Simply put, for a centrifugal device reducing the speed has a one-to-one effect on reducing airflow. Thus the *First Fan Law* states that if you were to reduce the speed of the centrifugal fan by 50% you would also reduce the amount of airflow by 50%.

The *Second Fan Law* states that a reduction in speed has the effect of reducing the system static pressure by the square of the reduction.

The *Third Fan Law*, which covers energy (horsepower), states that horsepower is reduced by the cube (^{3}) of the reduction in speed. So the fan law determines that a 20% reduction in fan speed results in about 50% reduction in consumer fan horsepower.

In most office buildings, with VAV systems operating in the 50% total flow range, you can see by examining the profile chart in Figure 9-16 just how effective this is at saving horsepower and consequently reducing energy costs.

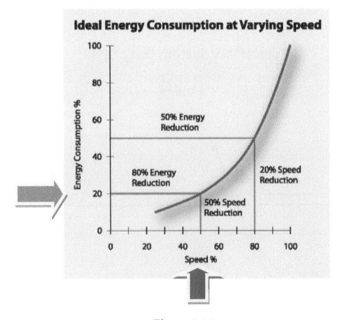

Figure 9-16

Another important result is that the reduction in fan speed also reduces duct leakage. This is a consequence of fan law number two whereby the system static pressure is reduced by the square of the reduction of airflow. By reducing the duct pressure by 30 percent when less air is required, almost instantaneous fan energy savings of more than 50 percent can be achieved above and beyond the application of a VFD.[41]

The prevalence of variable air volume systems has resulted in the transition from pneumatically controlled variable air volume to digitally controlled variable air volume terminal boxes. Traditionally, fan speed for a variable air volume system was controlled by installing a static pressure sensor approximately 2/3 of the distance down the main supply. Digital variable air volume boxes provide an important opportunity for additional savings since the position and airflow of each box is available in the buildings automation system. Current control strategy, often referred to as static pressure reset, reflects this capability by implementing the system to continually adjust the fan speed until only one of the variable air volume boxes is open 100%.

By definition, the static pressure reset method[42] of control should establish the minimal airflow needs of the facility since the worst zone is the one zone that is open 100%. Of course, code requirements for minimal airflow throughout the building will still benefit that must be considered. Keep in mind that, if both temperature and pressure resets are to be implemented simultaneously, some thought must be given to how these savings measures will interact.[43] See Figure 9-17.

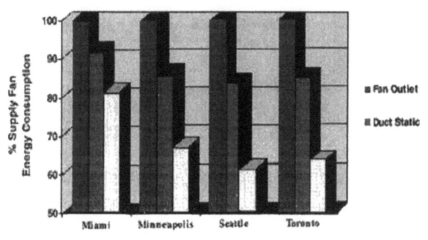

Figure 9-17

PUMPS

The vast majority of pumps encountered will have two things in common: they will be driven by an electrical motor and they will be a centrifugal device. Since centrifugal pumps follow the same laws as centrifugal fans, we have included the subject in this section.

Pumps can be a significant energy consumer in many applications and, since they are usually out of sight and out of mind until a failure occurs, consistent inspection and maintenance is essential. Opportunities for improvements in pumping energy consumption include addressing "throttle-valves" and their applications. Pumping systems tend to be oversized from a design standpoint for many valid reasons including the necessity to compensate for future changes and additions to pumped loops as well as the many additional system restrictions that occur during the construction process. It is typical in many buildings to discover that the pumping system has been adjusted or balanced by the use of a throttling valve, which is functionally the same as a discharge damper on a centrifugal fan. This actually results in increased horsepower while attempting to limit the flow. A motor is a load driven device and, up to the limits of its capability, its electrical consumption is a function of the load. One example of the savings potential from variable speed pumping is provided by Grundfos Pumps[44] for a high-rise office building in Los Angeles:

> Brookfield Properties performed before-and-after system monitoring of voltage and current as part of its energy audit. The "before" monitoring included a full week (24/7) of gathering data electronically from the building's existing pump system. The same monitoring was performed on the new system after sufficient time was allowed to tune the controls. On average, the old system used 5,000 kWh of electrical power per week, while the new system used 1,350 kWh per week. With an average cost of 11 cents per kilowatt-hour, the savings added up to more than $20,000 per year. The total equipment running hours dropped from an average of 184 hr per week for a six-pump system to 60 hr per week for the BoosterpaQ MPC. The installation was projected to save between $15,000 and $20,000, but during the first year alone it doubled projections with $40,000 in energy and maintenance cost savings.

One potential opportunity is to remove the throttling valves—or at least open them fully—and install a VFD for the purpose of adjusting the speed and flow of the pump. Of course having a VFD also allows for simple adjustments in the future.

It is also common to find pumped loops that run continuously especially where the consumers of the pumped fluids are controlled by three-way bypass valves. Earlier in this section we discussed the benefits of variable primary flow chilled water systems which results in substantially reduced pumping horsepower and the principle is exactly the same. One should focus on using the pumps both when needed and at the quantity needed and having the ability to adjust the pumping is essential.

The pumping system can be seen like a "batch" operation. Batch pumps should only be operated when a batch is being processed. Domestic water pumps in office buildings are frequently found to be running at constant speed even when a lesser or no flow condition exists.

Pumping systems tend to be significant users of energy that offer many opportunities for conservation and improvement, especially in industrial applications. The PSAT[45] tool, available at no charge from the DOE is a powerful tool that can be used to assess and improve pumping systems.

LIGHTING

Introduction

The design and application of electrical lighting is normally handled by lighting engineers and is governed by provisions generally furnished by ASHRAE and IES. ASHRAE 189.1-2011, the joint standard of ANSI/ASHRAE/USGBC/IES, and ASHRAE 90.1 provide us with numerous guidelines for applying high-performance to lighting. Advances in efficient lighting technology, lamp design, integration with daylighting, occupancy sensing, whole-building energy management systems and lighting control systems have all contributed to a significant increase in energy efficiency with lighting in both new buildings and retrofits in existing buildings. Some important facts from DOE's Energy Star Program:[46]

> *Lighting uses about 18 percent of the electricity generated in the U.S., and another 4 to 5 percent goes to remove the waste heat generated by those lights. Lighting in commercial buildings accounts for close to 71 percent of overall lighting electricity use in the U.S.* **With good design, lighting energy use in most**

buildings can be cut at least in half while maintaining or improving light-
ing quality. Lighting takes a larger share of a building's electricity use than any
other single end use—more than 35 percent.

In existing buildings, when it comes to energy retrofits, lighting is often lower in capital intensity, higher in ROI, and takes less time to pay for itself.[47] Given these facts, the potential impact for efficiency in lighting on electricity usage in the US can hardly be overestimated. See Figure 9-18

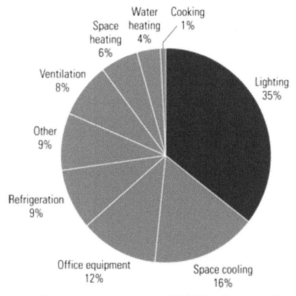

Courtesy: E SOURCE; data from *2005 Buildings Energy Data Book*

Figure 9-18

Terminology
Footcandles

The measurement of the light level on a surface (illuminance). The Illuminating Engineering Society provides standards which are widely adopted for the required footcandles based on various tasks that might be performed in a given area. These footcandle levels as well as other standards suggested by IES have been incorporated into ASHRAE 90.1 standards.

Some typical light levels:

Parking lot	2 Footcandles
Hallways	10 Footcandles
Factory floor	30 Footcandles
Offices	50 Footcandles
Inspection	100 Footcandles
Operating room	1000 Footcandles

Efficacy

A measure of the light output of the fixture in lumens compared to the watt inputs in electricity. The higher the efficacy of a fixture the more efficient it is at providing light, (NOTE: Some fixtures such as LED, have a different form of analysis instead of just efficacy). Tables 9-6 and 9-7 provided to illustrate the relative performance of various types of lamps. They are provided by Greenco Light and Power.[48]

Table 9-6.

Lamp type	Luminous efficacy (lumens/ watt)	Luminous output (lumens)	Wattage	CCT[a] (typical/ dominant wavelength)	CRI[b]	Lifetime
LED white package (cool)	132 lm/W	139 lm	1.05 W	6500 K	75	50k hours
LED white package (warm)	78 lm/W	87.4 lm	1.12 W	3150 K	80	50k hours
LED lamp (warm)	62 lm/W	650 lm	10.5 W	3000 K	92	50k hours
OLED panel	23 lm/W	15 lm	0.65 W	2800 K	75	5k hours
HID[c] (high watt) lamp system	120 lm/W 111 lm/W	37800 lm	315 W 341 W	3000 K	90	20k hours
Linear fluorescent lamp system	111 lm/W 97 lm/W	2890 lm 5220 lm	26 W 54 W	4100 K	85	25k hours
HID (low watt) lamp system	104 lm/W 97 lm/W	7300 IM	70 W 75 W	3000 K	90	12k hours
CFL	63 lm/W	950 lm	15 W	2700 K	82	12k hours
Halogen	20 lm/W	970 lm	48 W	2750 K	N/A	4k hours
Incandescent	15 lm/W	900 lm	60 W	3300 K	100	1k hours

Table 9-7.

POOR - **AVERAGE** - GOOD

Incandescent / Halogens	Mercury Vapor	Metal Halide	High Pressure Sodium	Compact Fluorescent (CFLs)	Light Emitting Diodes (LEDs)
Efficacy: 15-25 Lm/W	Efficacy: 30-60 Lm/W	Efficacy: 80-105 Lm/W	Efficacy: 70-110 Lm/W	Efficacy: 40-70 Lm/W	Efficacy: 60-140 Lm/W
CRI: **98-100**	CRI: **40-50**	CRI: **60-70**	CRI: **20-30**	CRI: **60-90**	CRI: **70-95**
Lifespan: **3,000 hours**	Lifespan: **15,000 hours**	Lifespan: **10,000 hours**	Lifespan: **15,000 hours**	Lifespan: **15,000 hours**	Lifespan: **40,000 hours**
- WATTAGE COMPARISON TABLE -					
40 - 60	15-25	5-15	5-15	12-15	5-8
60 - 75	25-35	15-25	15-25	15-18	7-10
75 - 100	35-45	20-35	20-35	18-23	10-15
100-150	50-60	25-40	25-40	23-35	15-20
150-200	70-85	35-45	35-45	30-45	20-25
200-250	90-110	40-55	40-55	45-60	25-30
Incandescent / Halogens	Mercury Vapor	Metal Halide	High Pressure Sodium	Compact Fluorescent (CFLs)	Light Emitting Diodes (LEDs)

Guidelines

The ASHRAE/IES 90.1 Standard, Section 9, gives power allowances for lighting applications in HP buildings. See Table 9-8.

In addition to specifying the lighting power density for interior lighting, exterior lighting power is also addressed by use type. Other important considerations for a high performance lighting system include the provisions for occupancy sensing with multi-level switching in hallways, dormitories, hotel and motels, commercial and industrial storage stack areas and library stacks. Lighting standards for egress and security restrict intensity to 0.1 watt/sq. ft. when continuous use is needed. Additional egress and security lighting is permitted as long as automatic occupancy sensing for shut off is implemented.

Table 9-8

ASHRAE/IES 90.1 Lighting Power Densities using the Building Area Method.

Building Type	Maximum Lighting Power Density (W/sq.ft.) Allowed Per Version of the ASHRAE/IES 90.1 Standard			
	1989	1999/2001	2004/2007	2010
Automotive Facility	0.96	1.5	0.9	0.982
Convention Center	2.07	1.4	1.2	1.08
Court House	1.44	1.4	1.2	1.05
Dining: Bar Lounge/Leisure	1.37	1.5	1.3	0.99
Dining: Cafeteria/Fast Food	1.37	1.8	1.4	0.90
Dining: Family	1.37	1.9	1.6	0.89
Dormitory	1.15	1.5	1.0	0.61
Exercise Center	2.07	1.4	1.0	0.88
Gymnasium	2.07	1.7	1.1	1.00
Healthcare Clinic	1.44	1.6	1.0	0.87
Hospital	1.44	1.6	1.2	1.21
Hotel	1.15	1.7	1.0	1.00
Library	1.29	1.5	1.3	1.18
Manufacturing Facility	0.96	2.2	1.3	1.11
Motel	1.15	2.0	1.0	0.88
Motion Picture Theater	2.07	1.6	1.2	0.83
Multi-Family	1.15	1.0	0.7	0.60
Museum	2.07	1.6	1.1	1.06
Office	1.26	1.3	1.0	0.90
Parking Garage	1.03	0.3	0.3	0.25
Penitentiary	1.44	1.2	1.0	0.97
Performing Arts Theatre	2.07	1.5	1.6	1.39
Police/Fire Station	1.44	1.3	1.0	0.96
Post Office	1.44	1.6	1.1	0.87
Religious Building	2.07	2.2	1.3	1.05
Retail	2.25	1.9	1.5	1.40
School/University	1.29	1.5	1.2	0.99
Sports Arena	2.07	1.5	1.1	0.78
Town Hall	1.44	1.4	1.1	0.92
Transportation	2.07	1.2	1.0	0.77
Warehouse	1.03	1.2	0.8	0.66
Workshop	0.96	1.7	1.4	1.20

Lighting Types

Incandescent

The oldest electric lighting technology. This class of lamps also includes halogen. Advantages: inexpensive, easier disposal. Disadvantages: relatively short life, low-efficiency, being phased out.

Compact Fluorescent Lights (CFL)

Advantages of being highly energy-efficient, long-lasting, and self-ballasted. Disadvantages: higher cost, higher disposal cost.

Fluorescent Lamps

The most common for commercial interiors in the United States. Some older buildings may still find some of the F40/T12 lamps and magnetic ballasts in use. However the latter rapidly being replaced by T8 or T5 lamps with electronic ballasts. As of July 1, 2010 it became illegal to manufacture or import T12 magnetic replacement ballasts and on July 14, 2012, standard T12 lamps are no longer being made. In addition to the commonly found 4' and 8' linear fluorescent, there are also "U" tube fluorescents in many 2' x 2' ceiling fixtures. These tubes are both expensive to operate and replace when using magnetic core and coil ballasts.

Building owners have two options for replacing T12 lamps; switching to T8s or, even better, T5s. These are more efficient lamps with lower mercury contents. Switching to T8s could cut mercury content by 43 percent and switching to T5s could cut mercury content by 56 percent.

A lamp that is labeled as F34T12/841 is fluorescent, 34 Watts, T12 indicating the thickness of the tube in 1/8" inch increments, so a T12 refers to 12 increments of 1/8" or 1 ½" in diameter (D). T8s are therefore 1" D and T5s are 5/8" D. In this example the /841 labeling adds two other pieces of information. "8" refers to the color rendering index of "80" which is a function of how well colors are perceived with the particular blend of phosphors in the tube. The higher the color rendering index number the better color can be perceived. Figure 9-19 is a color rendering chart showing some comparisons.

The "41" in the example refers to the color temperature in Kelvins, in this case, 4100 K, which can be seen as "cool white" in Figure 9-20.

High Intensity Discharge (HID)

Like fluorescent lights, these lamps function with ballasts and produce light by discharging an electric arc through a tube filled with gas. This type of lighting is found mostly in industrial applications usually with higher wattage outputs in the 400W to 1000W range. They are often found in high-bay uses such as loading docks, warehouses, as well as big box stores with occasional application in office and retail.

Typical Colour Rendering Index 0-100

	Typical CRI
Metal Halide 400W	65
High Pressure Sodium	22
T12 60W	70
T8 32W	78
T5 54W	85
Induction Lamp 200W	85

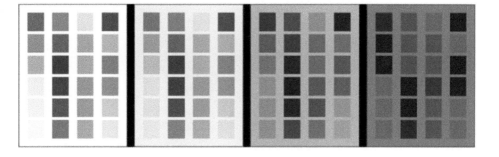

THE ABOVE CHART SHOWS HOW THE REPRODUCTION CO COLOUR IS AFFECTED BY BOTH THE CRI AND THE BRIGHTNESS OF A LIGHT SOURCE ILLUMINATING AN OBJECT.

Figure 9-19. Color Rendering Index (CRI)[49]

While they generally have long life (25,000 hours) and have reasonable efficacy, they do require 12-15 minutes warm up time before reaching full output which can be an issue in the event of a temporary power loss. HID lighting fixtures include metal halide, high-pressure sodium, low-pressure sodium and, although less common, mercury vapor.

Metal halide fixtures generally have good color rendition and reasonable lumen maintenance. Metal halide has traditionally been the correct choice for lighting large areas that require adequate color rendition.

High-pressure sodium (HPS) fixtures with higher efficacy than metal halide fixtures are generally good economic choices for outdoor use or for industrial applications where color rendition is not the driving factor. HPS is common in parking lots. Generally speaking, HPS is best used in areas where it is the only type of illumination, and is not being

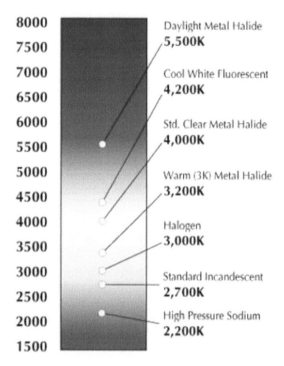

Figure 9-20. (Chart from mapawatt.com)[50]

compared with other types of lighting.

Low-pressure sodium has the highest efficacy of any commercially available HID, at the same time having the worst color rendition. Originally these lamps were very common in parking lot applications, although problems can arise with poor color rendition as an issue for security cameras.

Light-emitting Diode (LED)

LEDs are a solid state semi-conductor light source attached to a substrate which emit light when a current is applied to produce electroluminescence. LEDs are so efficient that they offer the potential to reduce the lighting costs in the USA by 50% and save $250 billion in energy over the next 20 years.[51] They have a very long life compared with other lamps[52]; the obstacle is first cost. See Figure 9-21.

Unlike traditional lamps, LEDs are an omni-directional light source, projecting light in a 360° pattern. This requires re-engineering

Figure 9-21

of the traditional light fixtures and reflectors to properly focus the useful light output from LEDs. LEDs have so far been relatively expensive compared even with other HP lamps, but they are falling in price and are starting to become a cost effective choice for a number of applications.

Traffic lights are a good example of the energy savings potential of the LED.[53]

Green 12" ball	140 W	13 W LED
Red 12" ball	140 W	11 W LED
Life	1 year	7 years LED
Cost	$3	$75 for LED

So, in this example, the LED uses about one tenth the power, lasts 7 times as long, but costs 25 times as much as the conventional lamp used in the traffic light. Obviously kWh costs, lumen depreciation (see Figure 9-22) and replacement costs need to be factored in to the calculation to assess the overall ROI. Because LEDs are a relatively new technology and they have such a long life, it hasn't been practical to test the entire expected life-cycle of the device. Philips has done a great deal of research on the technology and, along with DOE, has published a number of studies on life-span and expected depreciation.[54]

Lumen Depreciation

Another critical factor in lighting design is lumen depreciation, and Figure 9-23 illustrates the light output losses inherent in various type of lighting. It is important to consider the fact that lighting systems

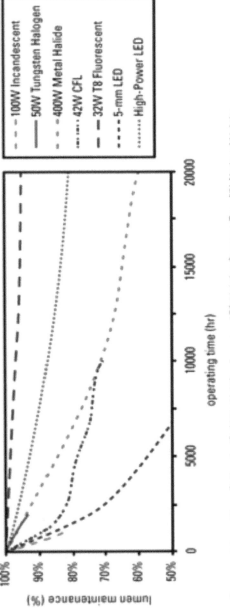

Figure 9-22

Source: Adapted from Bullough, JD. 2003. *Lighting Answers: LED Lighting Systems.* Troy, NY. National Lighting Product Information Program, Lighting Research Center, Rensselaer Polytechnic Institute.

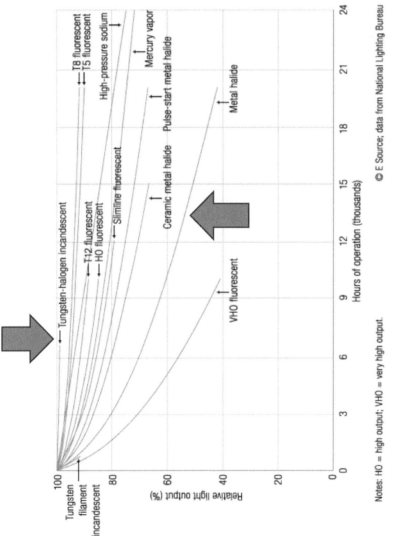

Figure 9-23

tend to be "over-designed" in order to compensate for the known loss of lumen output over the lifespan of the lamps. An understanding of this principle can be had by looking at the degradation curve for two common types, incandescent tungsten-halogen and metal halide. The incandescent curve shows essentially the same lumen output over the course of its relatively short 700-800 hour lifespan. A metal halide lamp, on the other hand, while enjoying a lifespan of about 25,000 hours has lost about 50% of its output at about 50% of its life.

Successful lighting application requires knowledge of numerous other factors which are not within the scope of the discussion here. Among these are visual comfort probability, room cavity ratios, coefficient of utilization, glare, uniformity and photometric considerations.[55]

High Performance Lighting Retrofits
The following is a series of case studies that propose lighting retrofits which take advantage of current technology, save energy, as well as improve occupant perceived lighting comfort.

Example #1—Upgrade T12 lighting to T8
A hospital had 415 T12 fluorescent fixtures, which operated 24 -7-365. The lamps and ballasts were replaced with T8 lamps and electronic ballasts, which saved about 30% of the energy and provided higher quality light. Although the T8 lamps cost a little more (resulting in additional lamp replacement costs), the energy savings quickly recovered the expense. In addition, because the T8 system produced less heat, air conditioning requirements during summer months are reduced. Conversely, heating requirements during winter months will be increased.

Calculations kW Savings
(# fixtures) [(Present input watts/fixture)- (Proposed input watts/fixture)]
(415)[(86 watts/T12 fixture)-(60 watts/T8 fixture)] = 10.8 kWh Savings
　　　= (kW savings)(Annual Operating Hours)
　　　= (10.8 kW)(8,760 hours/year)
　　　= 94,608 kWh/year

Air Conditioning Savings
(kW savings}(Air Conditioning Hours/year)(1/Air Conditioner's COP)
(10.8 kW}(2000 hours)(1/2.6)
8,308 kWh/year

Additional Gas Cost

(kW savings)(Heating Hours/year)(.003413 \MCF/kWh)(1/Heating Efficiency)(Gas Cost)

(10.8 kW)(1500 hours/year)(.00.3413 MCFI/kWh)(1/0.8)($4.00/MCF)

    = $276/year

Lamp Replacement Cost

    = [(# fixtures)(# lamps/fixture)!((annual operational hours/proposed lamp life)(proposed lamp cost))-((annual hours operation/present lamp life)(present lamp cost))]

[(415 fixtures)(2lamps/fixture)ll((8,760 hours/20,000 hours}($3.00/T8 lamp))--((8,760 hours/20,000 hours)($1.50/T12lamp))]

$545/year

Total Annual Dollar Savings

(kW Savings)(kW charge) [(kWh savings) + (Air Conditioning savings)](kWh cost) -(Additional gas cost)-(lamp replacement cost)

    = (10.8 kW)($120/kW year) + [(94,608 kWh) + (8,308kWh)]($0.05/kWh) -($276/year)-($545/year)

    = $5,621/year

Implementation Cost

    = (It fixtures) (Retrofit cost per fixture)

    = (415 fixtures)($45/fixture)

    = $18,675

Simple Payback

(Implementation Cost)/(Total Annual Dollar Savings) ($18,675)/($5,621/year)

3.3 years

Example #2—Replace Incandescent Lighting with
Compact Fluorescent Lamps

    A power plant has 111 incandescent fixtures that operate 24 hours/day, year round. The incandescent lamps were replaced with compact fluorescent lamps, which save over 70%, of the energy and last over ten times as long. Because the lamp life is so much longer, there is also a maintenance relamping labor savings. Air-conditioning savings or heating costs were not included, because these fixtures are located in a high-bay building that is not heated or air-conditioned.

Calculations

Watts Saved Per Fixture

(Present input watts/fixture)-(Proposed input watts/fixture)

(150 watts/fixture)-(30 watts/fixture) 120 watts saved/fixture

kW Savings

(it fixtures)(watts saved/fixture)(1 kW/1000 watts) (111 fixtures)(120 watts/fixture)(l/1000)

13.3kW

kWh Savings

(Demand savings)(annual operating hours)

(13.3 kW)(8,760 hours/year) 116,683 kWh/year

Lamp Replacement Cost

[(Number of Fixtures)(cost per CFL Lamp)(operating hours/lamp life)]-[(Number of existing incandescent bulbs)(cost per bulb)(operating hours/lamp life)]

- [(111 Fixtures)($10/CFL lamp)(8,760 hours/10,000 hours}] - ((111 bulbs)($1.93/type "A" lamp)(8,760 hours/750 hours))

$–1,530/year* ***Negative cost indicates savings.**

Maintenance Relamping Labor Savings

= [(# fixtures)(maintenance relamping cost per fixture)) (((annual hours operation/present lamp life))-((annual hours operation/proposed lamp life))] [(111 fixtures)\{$1.7/fixture))[((8,760/750}} ({8,760/10,000)))

= **$2,039/year**

Total Annual Dollar Savings

= (kWh savings)(kWh cost)+ (kW savings)(kW cost)

- (lamp replacement cost) + (maintenance relamping labor savings)

= (116,683 kWh)($.05/kWh)+(13.3)($120/kW year)

(-1,530I year) + (2,039I year)

= **$10,999/year**

Total Implementation Cost

[(# fixtures)(cost/CFL ballast and lamp))+ (retrofit labor cost)]

(111 fixtures)($45/fixture)

$4,995

Simple Payback
(Total Implementation Cost)/(Total Annual Dollar Savings)
($4,995)/(10,999/year)
0.5 years

Example #3—Install Occupancy Sensors
    In this example, an office building has many individual offices that are only used during portions of the day. After mounting wall-switch occupancy sensors, the sensitivity and time delay settings were adjusted to optimize the system. The following analysis is based on an average time savings of 35% per room. Air conditioning costs and demand charges would likely be reduced; however, these savings are not included.

Calculations
kWh Savings
{# rooms)(# fixtures/room)(input watts/fixture) (1 kW/1000 watts) (Total annual operating hours)(estimated % time saved/100)
(50 rooms)(4 fixtures/room)(144 watts/fixture)(1/1000)
(4,000 hours/year)(.35)
    = 40,320 kWh/year

Total Annual Dollar Savings ($/Year)
    = (kWh savings/year)(kWh cost)
    = (40,320kWh/year)($.05/kWh)
    = $2,016/year

Implementation Cost
    = (# occupancy sensors needed)[(cost of occupancy sensor)+(installation time/room)(labor cost)]= (50)[($75)+(1 hour/sensor)($20/hour)]
    = $4,750

Simple Payback
    = (Implementation Cost)/(Total Annual Dollar Savings)
($4,750)/($2,016/year)
    = 2.4 years

Example #4—Retrofit Exit Signs with LEDs
    An office building had 117 exit signs that used incandescent bulbs. The exit signs were retrofitted with LED exit kits, which saved 90% of

the energy. Even though the existing incandescent bulbs were "long-life" models, which are expensive, material and maintenance savings were significant. Basically hospital should not have to relamp exit signs for 10 years.

Calculations
input Wattage- Incandescent Signs
　　　= (Watt/fixture) (number of fixtures)
　　　= (40Watts/fix)(117 fix)
　　　= 4.68kW
Input Wattage -LED Signs
　　　= (Watt/fixture) (number of fixtures)
　　　= (3.6 Watts/fix) {117 fix)
　　　= .421 kW

kW Savings
　　　= (Incandescent Wattage) - (LED Wattage)
　　　= (4.68 kW) - (.421 kW)
　　　= 4.26 kW

kWh Savings
　　　= (kW Savings)(operating hours)
　　　= (4.26 kW)(8,760 hours)
　　　= 37,318 kWh/yr.

Material Replacement Cost
((Number of LED Exit Fixtures)(cost per LED Fixture)(operating hours/Fixture life)] - ((Number of existing Exit lamps)(cost per Exit lamp){operating hours/lamp life)]
[(117 Fixtures) ($60/lamp kit)(8,760 hours/87600 hours)] –
[(234 Exit lamps))($3.00/lamp)(8,760 hours/8,760 hours)]
　　　= ($702.00 - $702.00 = $0.00)/year

Maintenance Relamping Labor Savings
　　　= (# signs)(Number of times each fixture is relamped/yr.)(time to relamp one fixture)(Labor Cost)
　　　= (117 signs)(l relamp/yr)(.25 hours/sign)($20/hour)
　　　= $585/year

Annual Dollar Savings
　　　= [(kWh savings)(electrical consumption cost)] + [(kW savings)(W-cost)] + [Maintenance Cost Savings]
　　　= [(37,318 kWh)($.05/kWh)] + [(4.26 kW)($120/kW yr)]+ [$585/yr]
　　　= $2,962/year

Implementation Cost
> = # Proposed Fixtures]((Cost/fixture+ Installation Cost/fixture)]
> = [117][$60/fixture + $5/fixture]
> = $7,605

Simple Payback
> = (Implementation Cost)/(Annual Dollar Savings)
> = ($7,605)/($2,962/yr)
> = 2.6 years

Example #5—Replace Outdoor HID lighting with LEDs

A great opportunity exists for the application of LEDs with outdoor lighting. LEDs produce very "white" light, and the human eye "perceives" that white light is brighter under relatively low-light (nighttime) conditions. Figure 9-24 and Table 9-9 demonstrate this possible application.

In other words, "photopic" footcandles are measured with a light meter, whereas "scotopic" footcandles are a "perceived" factor. One thing to point out is how well the LEDs provide "uniformity" of lighting in the parking lot compared to the "hot spots" directly under the sodium lighting on the right side of the photo. This factor contributes importantly to the perceived brightness of the LED side.

Example #6—replace "U" tube Lamps with T5 Lamps

The existing fixtures were 2' by 2' lay-in troffers with two F40T12CW "U" lamps and a standard ballast consuming 96 watts per fixture. The retrofit was to remove the "U" lamps and install three F14T5 lamps with electronic ballasts, which had only 32 watts per fixture while at the same time providing 15% increase in footcandles.

Emerging Technologies
Linear LEDs

These are for replacement of T12 or possibly T8 lamps in office type luminaires and are offered by several manufacturers. As of this writing they are an expensive choice when compared to T8 or T5 lamps with electronic ballasts. Linear LEDs come in two general varieties: the non-dimmable type has the LED power supplies ("drivers") built in to the tube eliminating a separate power conditioner; the second type is dimmable and requires an external power supply.

One advantage of the linear LEDs compared with conventional

Figure 9-24

Table 9-9. Parking Lot Calculations—Scotopic Considerations

Item	HPS	LED
Total System Wattage	300 W	141 W
Average Delivered Lumens per fixture (photopic)	19,000	8,040
Average Footcandles (photopic)	1.96	1.01
Average Delivered Lumens (scotopic)	11,780	17,206
Average Footcandles (scotopic)	1.22	2.16

Photopic vision is how the eye perceives objects and colors under bright light. Conversely, scotopic vision is how the eye perceives objects and colors under low-light conditions, such as a parking lot at night. The above measurements show that LED lights provide more perceived light at night while using much less energy.

tube lamps is that all of the output is highly directional. The amount of lighting output wasted bouncing around in the typical fluorescent fixture is drastically reduced benefiting the efficiency calculations. At this time linear LEDs are quite expensive—as much as $90 per tube—when compared to a T8 at around $2.00.

Digital Addressable Lighting Interface (DALI)

The International Electrotechnical Commission[57] has developed Standards which are part of the DALI and which contain automation protocols governing network interfaces for lighting in building automation.

Electronic ballasts for both fluorescent and HID lighting applications may be acquired from numerous manufacturers as "network" items. As such these ballasts have unique "addresses" which permits them to be interfaced with several types of building automation systems. When used with a wireless control system, these ballasts can be controlled on an individual basis both for purposes of dimming or turning on and off completely. The opportunity to individually control each lighting fixture provides significant opportunity to save energy when coupled with an intelligent automation and control system.

Lighting Control Systems

Lighting control systems are one of the most effective ways to increase lighting efficiency in buildings. They include: occupancy sensors, dimmers, time scheduling, bi-level switching (switching down for less intensive tasks), automatic daylight dimming (sensing daylight levels and reducing electrical lighting accordingly) and demand limiting (cutting lighting wherever possible at peak demand), among other applications. Automatic lighting control systems can be centralized intelligence, with a central processor, or distributed intelligence, where each lighting device has its own processor, depending on the requirements of the application. The latter provides more flexibility and avoids system failure. Networked systems with wireless controls in each fixture are superior in design and are the future of all control systems. An Addendum to ANSI/ASHRAE/IES Standard 90.1-2010, Energy Standard for Buildings Except Low-Rise Residential Buildings, which sets minimum guidelines for automatic lighting control systems, has been proposed in 2013 to clarify and expand the format for assessments. The FEMP and the LBNL/ORNL publish studies and guides which provide detailed information on how to implement a variety of strategies and calculate potential energy savings.[58]

Table 9-12, published by the WBDG, is a very useful and practical guide for lighting control system applications.[59]

Occupancy Sensing

Occupancy sensors can be retrofitted into individual spaces to provide local control but can also be networked into building automation systems. Intelligent occupancy sensing (IOS) systems were developed to sense occupants in areas which are intermittently occupied and where a high degree of overall control is required. Figure 9-25 is an example system developed by Texas Instruments.[60]

Table 9-10

Typical Lighting Control Applications			
Type of Control	**Private Office**	**Open Office - Daylit**	**Open Office - Interior**
Occupancy Sensors	++	++	++
Time Scheduling	+	++	++
Daylight Dimming	++	++	0
Bi-Level Switching	++	+	+
Demand Lighting	+	++	++

++ = good savings potential

+ = some savings potential

0 = not applicable

Dimming

Often overlooked as an energy savings option, dimmers can save as much as 19% in electricity costs per fixture.[61] Automatic daylight dimming can further increase savings up to 30% depending on the daylight available in the specific building.[62]

Table 9-11

	Operating Cost Comparison Open Office Area, 1000 sq. ft.				
Performance	Base Case	Time Scheduling	Occupancy Sensors	Daylighting	Time Scheduling + Daylighting
Annual Energy Use[a]	5700 kWh	5100 kWh	5000 kWh	4200 kWh	3700 kWh
Annual Energy Cost	$340	$305	$300	$250	$220
Annual Energy Cost Savings	—	$35	$40	$90	$120

[a] Average daily "on" hours for wall switch is 9.1. Average daily occupied hours for the office is 6.8.

Wireless

A number of manufacturers produce lighting controls that operate with the Zigbee[63] protocol. This enables integrated building control without the added expense of extra control wiring and interfaces. It also enables the lighting systems to become a resource for demand response load systems, permitting reduced utility charges or actual cash rebates from the utility for participation.[64]

CONCLUSIONS

It is clear that major opportunities exist for energy and cost savings in individual buildings and the buildings sector as a whole by implementing high-performance measures in mechanical and electrical equipment. The systematic and integrated analysis of all related systems yields large benefits which can be accurately quantified and consistently achieved with continuous monitoring. Substantial, long-term savings need not require major capital outlays, and equipment retrofits can have ROIs ranging anywhere from 6 months to 5 years.

Table 9-12

Space Type	Typical Use Pattern	If...	Then...
Cafeterias or Lunchrooms	Occasionally occupied	Daylighted	Consider daylight-driven dimming or on/off control
		Occupied occasionally	Consider ceiling-mounted occupancy sensor(s). Make sure minor motion will be detected in all desired locations.
Classroom	Usually occupied	Multiple tasks like overhead projectors, chalkboard, student note taking and reading, class demonstrations	Consider manual dimming
	Occasionally occupied	Occupied by different groups of students and teachers daily	Consider ceiling- or wall-mounted occupancy sensor(s) and manual dimming. Make sure that minor motion will be detected.
		Lights left on after hours	Consider centralized controls and/or occupancy sensors.
Computer Room	Usually unoccupied	Lights are left on all the time	Consider occupancy sensors with manual dimming. Be sure that minor motion will be detected and that equipment vibration will not falsely trigger the sensor.
Conference Room	Occasionally occupied	Multi-tasks from video-conferencing to presentations	Consider manual dimming (possibly preset scene control)
		Small conference room	Consider a wall box occupancy sensor
		Large conference room	Consider ceiling- or wall-mounted occupancy sensor(s). Be sure that minor motion will be detected in all desired locations.
Gymnasium or Fitness	Usually occupied	Requires varied lighting levels for activities	Consider manual dimming and occupancy sensors. Be sure that the HVAC system will not falsely trigger the sensor.
	Occasionally occupied	Requires varied lighting levels for activities	Consider ceiling- and wall-mounted passive infrared occupancy sensors. Be sure that the coverage areas of the sensors are sufficiently overlapped to keep the lights on when the room is occupied.
Hallways	Any	Occasionally or usually occupied	Consider occupancy sensors with elongated throw. Be sure that coverage does not extend beyond the desired area.
		Daylighted	Consider daylight on/off control.

(Continued)

Table 9-12 (*Cont'd*)

Health Care— Examination Rooms	Occasionally occupied	Different lighting needs for examination	Consider manual dimming.
		Small areas	Consider a wall box occupancy sensor.
Health Care— Hallways	Usually occupied	Daylighted	Consider automatic daylight-driven dimming.
		Requires lower lighting level at night	Consider centralized controls to lower lighting levels at night.
Health Care— Patient Rooms	Usually occupied	Different lighting needs for watching television, reading, sleeping, and examination	Consider manual dimming. Occupancy sensors may not be appropriate.
Hotel Rooms	Occasionally occupied	Used primarily in the late afternoon through evening for sleeping and relaxing	Consider manual dimming.
Laboratories	Usually occupied	Daylighted	Consider automatic daylight-driven dimming in combination with occupancy sensors.
Laundry Rooms	Occasionally occupied	Requires high light levels, yet lights are usually left on	Consider occupancy sensors.
Libraries— Reading Areas	Usually occupied	Daylighted	Consider automatic daylight-driven dimming. Occupancy sensors may be appropriate.
		Lights left on after hours	Consider centralized controls.
Libraries—Stack Areas	Occasionally occupied	Stacks are usually unoccupied	Consider ceiling-mounted sensor(s).
Lobby or Atrium	Usually occupied but no one "owns" the space	Daylighted and lights should always appear on	Consider automatic daylight-driven dimming.
		It isn't a problem if lights go completely off in high daylight	Consider automatic daylight-driven dimming or on/off control.
		Lights are left on all night long, even when no one is in the area for long periods	Consider occupancy sensors. Be sure that minor motion will be detected in all desired areas.
Office, Open	Usually occupied	Daylighted	Consider automatic daylight-driven dimming.
		Varied tasks from computer usage to reading	Consider manual dimming.
		Lights left on after hours	Consider centralized controls and/or occupancy sensors.
Office, Private	Primarily one person, coming and going	Daylighted	Consider manual dimming, automatic daylight- driven dimming, or automatic on/off.
		Occupants are likely to leave lights on and occupants would be	Consider a wall box occupancy sensor. Add dimming capabilities if appropriate.

(Continued)

Table 9-12 (*Concluded*)

		in direct view of a wall box sensor	
		Occupants are likely to leave lights on and partitions or objects could hide an occupant from the sensor	Consider a ceiling- or wall-mounted occupancy sensor. Add dimming capabilities if appropriate.
Photocopying, Sorting, Assembling	Occasionally occupied	Lights are left on when they are not needed	Consider an occupancy sensor. Be sure that machine vibration will not falsely trigger the sensor.
Restaurant	Usually occupied	Daylighted	Consider automatic daylight-driven dimming.
		Requires different lighting levels throughout the day	Consider manual dimming (possibly preset scene dimming).
		Requires different lighting levels for cleaning	Consider centralized control.
Restroom	Any	Has stalls	Consider a ceiling-mounted ultrasonic occupancy sensor for full coverage.
		Single toilet (no partitions)	Consider a wall switch occupancy sensor.
Retail Store	Usually occupied	Daylighted	Consider automatic daylight-driven dimming
		Different lighting needs for retail sales, stocking, cleaning	Consider centralized controls or preset scene dimming control.
Warehouse	Aisles are usually unoccupied	Daylighted	Consider daylight-driven dimming or daylight on/off control.
		Lights in an aisle can be turned off when the aisle is unoccupied	Consider ceiling-mounted occupancy sensors with elongated throw. Select a sensor that will not detect motion in neighboring aisles, even when shelves are lightly loaded.

Improvements in the thermal and daylighting efficiency of the envelope will reduce the upfront requirements in new buildings for M&E equipment and can reduce the replacement costs in retrofits. Continuous commissioning and improvements in the predictive accuracy of building energy analysis programs will further accelerate the development of high-performance for HVAC and electrical lighting. As building automation systems and wireless networks increase in sophistication and become lower in cost, building owners and managers will have more powerful tools to track performance and insure results.

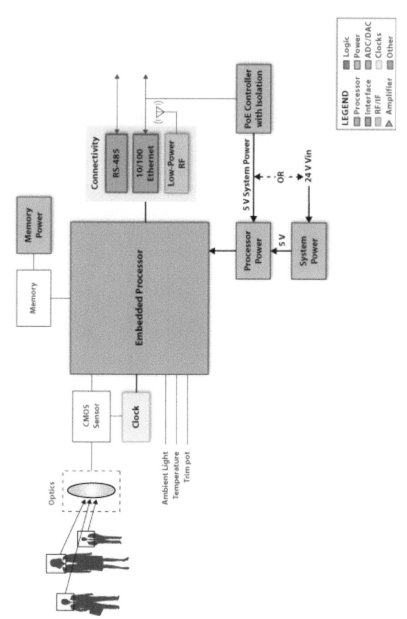

Figure 9-25

References

1. Residential, Commercial and Industrial.
2. SEE: http://buildingsdatabook.eren.doe.gov/ChapterIntro1.aspx?1#1, Section 1.1 Buildings Sector Energy Consumption, 1.19, 1.10, 1.12., 2010 figures.
3. ANSI/ASHRAE/USGBC/IES Standard 189.1-2011, Standard for the Design of High Performance Buildings
4. SEE: http://www.iesna.org/, SEE Also: https://www1.eere.energy.gov/buildings/ssl/partnership_ies.html
5. For a good general discussion, SEE: http://www.facilitiesnet.com/hvac/article/Condensing-Boilers-Understanding-Efficiency-Ratings--12041#
6. "Turndown is defined as the ratio of maximum fuel input rate to the minimum fuel input rate of a variable input burner. In general, it refers to the ability to adjust a boiler or water heater's firing rate (input) to precisely meet the real-time heating demand (output) of the system without temperature overshoot or wasteful cycling." SEE: http://www.deppmann.com/2009/10/19th/
7. SEE: http://www.peerlessboilers.com/Products/ResidentialBoilers/Peerless-PUREFIRE/tabid/196/Default.aspx
8. CASE STUDY performed by: Varo Engineers, Municipal Energy Program, City of Bay Village, Cuyahoga County, Ohio, http://www.varoengineers.com/
9. ECO: Energy Conservation Opportunity
10. "Specifically, AFUE is the ratio of annual heat output of the furnace or boiler compared to the total annual fossil fuel energy consumed by a furnace or boiler." SEE: http://energy.gov/energysaver/articles/furnaces-and-boilers
11. SEE: http://www.mnshi.umn.edu/definitions/FEMP_article_furnace_efficiency.pdf
12. SEE ALSO: http://efficiency.lbl.gov/drupal.files/ees/Electricity%20and%20Natural%20Gas%20Efficiency%20Imrovements%20for%20Residential%20Gas%20Gurnaces%20in%20the%20US_LBNL-59745.pdf
13. FOR A CASE STUDY WITH RADIANT HEATING PANELS, SEE: http://www.electricheat.com/Portals/0/Riveredge.pdf
14. SEE: http://www.ornl.gov/info/ornlreview/rev28_2/text/gas.htm
15. SEE: http://www.robur.com/products/pro-solutions/pro-gahp-line-ar-rtar-series/description.html
16. SEE: http://energy.gov/energysaver/articles/heat-pump-water-heaters, and diagram next page
17. Depending on latitude, below-ground temperatures vary from 45° - 75°F.
18. http://www.usgbcnv.org/Resources/Documents/Programs/Munch%20and%20Learn/GBCI%20CE%20presentation%20extended.pdf
19. SEE DIAGRAMS to follow: http://energy.gov/energysaver/articles/geothermal-heat-pumps
20. SEE: http://www1.eere.energy.gov/femp/pdfs/ghptf.pdf, and: http://www.geoexchange.org/index.php?option=com_content&view=article&id=48:geothermal-heat-pumps&catid=375&Itemid=32, also: http://www.energystar.gov/index.cfm?fuseaction=find_a_product.showProductGroup&pgw_code=HP
21. DCV, VAV and VFD are discussed later in the Chapter.
22. SEE: http://www1.eere.energy.gov/buildings/commercial/bba_rtu_spec.html
23. "A single-number figure of merit expressing cooling part-load EER efficiency for commercial unitary air conditioning and heat pump equipment on the basis of weighted operation at various load capacities for the equipment.integrated part-load value (IPLV)," SEE: http://wiki.ashrae.org/index.php/Integrated_energy_efficiency_ratio_(IEER), and : High Performance Rooftop Units, Howett, D.,

ORNL, 2011, http://www1.eere.energy.gov/femp/pdfs/ntwg_122011_howett.
pdf
24. SEE: http://www.daikinmcquay.com/main/McQuay/ProductInformation/Roof-
topSystems/Rebel
25. SEE: http://www.ahrinet.org/hvacr+industry+standards.aspx
26. IPLV: "The Integrated Part Load Value (IPLV) is a performance characteristic
developed by the Air-Conditioning, Heating and Refrigeration Institute (AHRI).
It is most commonly used to describe the performance of a chiller capable of
capacity modulation. Unlike an EER (Energy Efficiency Ratio) or COP (coefficient
of performance), which describes the efficiency at full load conditions, the IPLV is
derived from the equipment efficiency while operating at various capacities." See:
http://ashrae-cfl.org/2010/03/understanding-iplvnplv/
27. SEE: Carnegie-Mellon IWESS: http://www.cmu.edu/iwess/components/steam_
absorption_chiller/flow-diagram.html
28. SEE: http://www.gsa.gov/portal/content/140963, and: http://www1.eere.
energy.gov/femp/technologies/newtechnologies_techbrief.html
29. SEE: http://www.aft.com/documents/AFT-TME-Case-Study.pdf
30. SEE: ASHRAE Standard 62.1 -2010: Ventilation for Acceptable Indoor Air Quality
31. SEE: http://www1.eere.energy.gov/femp/pdfs/fta_co2.pdf
32. SEE: Understanding & Designing Outdoor Air Systems (DOAS), Mumma, S.,
ASHRAE Short Course, 2010, http://www.caee.utexas.edu/prof/Novoselac/
classes/ARE389H/Handouts/doas_arkansas_final.pdf
33. SEE: Underfloor Air Technology, http://www.cbe.berkeley.edu/underfloorair/
Introduction.htm, and, HVAC: The Challenges and Benefits of Under Floor Air
Distribution Systems, Spinazzola, S., facilitiesnet.com, 2005, http://www.facili-
tiesnet.com/hvac/article/Air-Distribution-Turned-Upside-Down--3516#
34. SEE: http://www.taylor-engineering.com/downloads/articles/ASHRAE%20
Journal%20-%20Economizer%20High%20Limit%20Devices%20and%20Why%20
Enthalpy%20Economizers%20Don't%20Work%20-%20Taylor%20&%20Cheng.pdf
35. Graphic above and explanation of dry-bulb changeover from: http://www.
advancedbuildings.net/files/advancebuildings/OSAeconoTechBrief05g_0.pdf,
Or SEE: Practical Controls: A Guide to Mechanical Systems, Calabrese, S., CRC/
Fairmont Press, 2003, http://books.google.com/books?id=tBm8aWJk06kC&pg=P
A208&lpg=PA208&dq=dry+bulb+changeover&source=bl&ots=HoqtvDh5fS&sig=
jh-POq3YQbhGdnfx6sLL8rvm6bA&hl=en&sa=X&ei=PjGYUc_TIu630QGLwYHw
Bw&ved=0CEkQ6AEwBg#v=onepage&q=dry%20bulb%20changeover&f=false
36. SEE: Introduction to Premium Efficiency Motors, Copper Development Association, 2013, http://www.copper.org/environment/sustainable-energy/electric-
motors/education/motor_text.html
37. "MotorMaster+ is a free online National Electrical Manufacturers Association
(NEMA) Premium® efficiency motor selection and management tool that supports motor and motor systems planning by identifying the most efficient action
for a given repair or motor purchase decision. The tool includes a catalog of more
than 20,000 low-voltage induction motors, and features motor inventory management tools, maintenance log tracking, efficiency analysis, savings evaluation,
energy accounting, and environmental reporting capabilities." http://www1.eere.
energy.gov/manufacturing/tech_assistance/software_motormaster.html
38. SEE: http://aceee.org/press/2012/08/nema-and-energy-efficiency-advocates
39. SEE: http://www.worldenergyoutlook.org/publications/weo-2012/
40. SEE: http://www1.eere.energy.gov/manufacturing/tech_assistance/pdfs/re-
place_vbelts_motor_systemts5.pdf

41. ASHRAE Transactions January 1, 2010 | Liu, Mingsheng; Feng, Jingjuan; Wang, Zhan; Wu, Lixia; Zheng, Keke; Pang, Xiufeng |, http://www.thefreelibrary.com/Impacts+of+static+pressure+reset+on+VAV+system+air+leakage,+fan...-a0227975411

42. SEE also: Increasing Efficiency with VAV System Static Pressure Setpoint Reset, Taylor, S. T., P.E., ASHRAE Fellow, ASHRAE Journal, June 2007. http://www.taylor-engineering.com/downloads/articles/ASHRAE%20Journal%20-%20Supply%20Pressure%20Reset.pdf

43. For a thorough explanation and discussion of various types of optimized Static Pressure Control for VAV Systems see: http://www.trane.com/commercial/uploads/pdf/866/VentilationFanPressureOptimization.pdf

44. http://energy.grundfos.com/en/pump-audit/pump-audit-cases/building-service/pump-replacement-saves-$40000-annually-in-energy-and-maintenance

45. SEE: http://www1.eere.energy.gov/manufacturing/tech_assistance/software_psat.html

46. SEE: http://www.energystar.gov/index.cfm?c=business.EPA_BUM_CH6_Lighting

47. SEE: Better Buildings Alliance (BBA) Lighting and Electrical Project Team, http://www1.eere.energy.gov/buildings/commercial/bba_lighting_team.html And: http://www1.eere.energy.gov/buildings/betterbuildings/bba/bba-index.html

48. http://greencolightandpower.com/main/images/chart.jpg

49. SEE: http://www.gilus.us/Elements/CRI%20Chart.jpg

50. SEE: http://mapawatt.com/wp-content/uploads/2010/06/color_temperature_charts.jpg

51. SEE: http://www1.eere.energy.gov/buildings/ssl/sslbasics_ledbasics.html

52. Estimates range from an average of 35K - 50K hours (DOE) (compared with, e.g., 10K hours with CFLs) but there is some debate about this and usage factors have to be examined carefully during assessment.

53. SEE: http://web.archive.org/web/20090505080533/http://www1.eere.energy.gov/buildings/ssl/comparing.html

54. SEE: Evaluating the Lifetime Behavior of LED Systems, Phillips LumiLEDS, available at: www.philipslumileds.com/uploads/167/wpf-15pdf, and Lifetime of White LEDs: Building Technology Program, DOE-EERE, available at: http://apps1.eere.energy.gov/buildings/publications/pdfs/ssl/lifetime_white_leds.pdf

55. A good general Reference for lighting terminology and definitions can be found at the Lighting Research Center: http://www.lrc.rpi.edu/programs/NLPIP/glossary.asp, Also, IES's Nomenclature and Definitions for Illuminating Engineering, http://www.ies.org/store/product/nomenclature-and-definitions-for-illuminating-engineeringbr-rp1605-1013.cfm

56. SEE: http://www.cree.com/lighting/products/outdoor

57. IEC: http://www.iec.ch/about/?ref=menu

58. For Example: A Meta-Analysis of Energy Savings from Lighting Controls in Commercial Buildings, Williams, A., Atkinson, B., Garbesi, K., Rubinstein, F., Energy Analysis Department, Lawrence Berkeley National Laboratory, September 2011, http://efficiency.lbl.gov/drupal.files/ees/Lighting%20Controls%20in%20Commercial%20Buildings_LBNL-5095-E.pdf, and: http://www1.eere.energy.gov/femp/technologies/eep_light_controls.html#demand

59. "With the move towards a Green environment, intelligent occupancy sensors offer a means to accomplish this task as significant energy savings from using such devices can be achieved; 40% to 46% in classrooms, 13% to 50% in private offices, 30% to 90% in restrooms, 22% to 65% in conference rooms, 30% to 80% in

corridors, and 45% to 80% in storage areas. Besides minimizing energy consumption, occupancy sensors can also be used for security (by indicating that an area is occupied.)" SEE: http://www.ti.com/solution/intelligent_occupancy_sensing#

60. http://www.lutron.com/en-US/education-training/Pages/tools/energysaving-calc.aspx

61. SEE: FEMP, Daylight Dimming Controls: http://www.rileyelectricalsupply.com/pdfs/9.pdf

62. ZIGBEE ALLIANCE: "ZigBee Building Automation offers a global standard for interoperable products enabling the secure and reliable monitoring and control of commercial building systems. It is the only BACnet® approved wireless mesh network standard for commercial buildings. Owners, operators and tenants can benefit from increased energy savings and ensure the lowest lifecycle costs with this green and easy-to-install robust wireless network. By using ZigBee Building Automation products in your building, you can contribute toward satisfying credits in the categories of Sustainable Sites, Energy and Atmosphere, Indoor Environmental Quality under the U.S. Green Building Council's LEED® green building certification program." http://www.zigbee.org/Standards/ZigBeeBuildingAutomation/Overview.aspx

63. SEE: http://www.lutron.com/en-US/Residential-Commercial-Solutions/Pages/Commercial-Solutions/WholeBuildingSolutions.aspx, also: http://www.leviton.com/OA_HTML/SectionDisplay.jsp?section=37663

Chapter 10

Operations & Maintenance

INTRODUCTION

A high-performance (HP) building is by definition a building with substantially lower operating costs and maintenance requirements. The design of a new HP building must have long-term waste reduction, energy and resource efficiency and lower environmental impact as part of the model. And as we have seen, validation must occur down the road with continuous feedback from building owners and facilities managers. The same must be true in an existing building which is retrofitted with the goals of high performance. Getting continuous feedback from monitoring systems that track weather, energy, water usage, materials and waste processing in building functions is essential, but operational behavior including occupant education is equally important to infrastructure improvements to achieve consistent results. Occupying and operating a high performance building is therefore a learning experience as well as a validation of computer modeling, engineering prowess and architectural mastery. After all, the ultimate goal of HP is net-zero energy and zero waste, and building it and maintaining it cost-effectively are a tremendous challenge. After it is built or retrofitted, the high-performance building is not a static structure which we just respond to. The building becomes a "living system" which is operationally dynamic and educational.

GENERAL PRINCIPLES

There are numerous strategies and frameworks for O&M programs but we can first consider some basic principles common to all of them before looking at specific examples and how they dovetail with HP buildings. The overall goal is to go from a building which is *high-cycling* to a building which is *low-cycling*. That is a building which consumes less of everything and therefore has less to cycle through every phase of its operation, occupancy and waste processing. We can extend the concept of "less is more,"[1] to the concept of *resource flow (RF)*, which is the cycling of all energy, water and materials (food, paper, chemicals, etc.), into, through, out of, and back into the site. All O&M programs which are high-performance then must consider the following with regard to *resource flow*:

Reduce Inputs

Effectively manage resource flow into the building: purchasing, acquisition and transport.

Economize Processes

Optimize resource flow through envelope, mechanical and electrical infrastructure. Coordinate and facilitate the operational behavior of the occupants. Reduce or eliminate single-pass operating systems.

Reduce Outputs

Minimize energy, water and material (RF) waste outputs. Maximize reuse and redistribution of waste outputs on-site or to other local users. Reduce or eliminate single-use consumption patterns.

Considering the above principles, we can provide a general schema for the O&M of a high-performance building, which extends the concept of continuous commissioning explained in Chapter 5, and which engages a continuous cycle as shown in Figure 10-1.

Another way to look at O&M and resource flow from the standpoint of life-cycle analysis (LCA), is shown in Figure 10-2.[2]

Every time we bring resources into the building, there are consequences from the standpoint of how the energy, water and materials impact the delivery systems, the building environment, and the environment external to the building. So the O&M of high-performance buildings has to take into account that the system boundaries of high-performance extend well beyond the building itself.

O&M PROGRAMS

We have seen in Chapter 3 that many standards, codes and rating systems have prescriptive paths for O&M procedures that can help meet

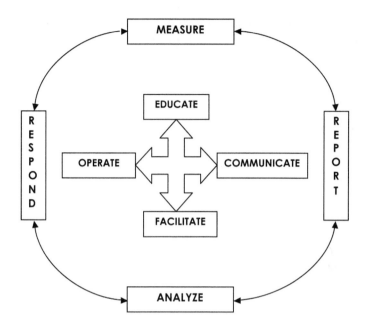

Figure 10-1.

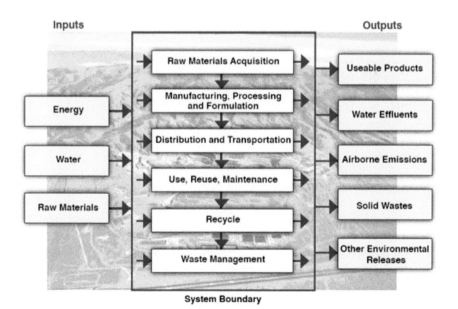

Figure 10-2.

the goals of high-performance. Most of these are designated procedures for new buildings and are part of the validation process for energy performance, resource management targets and environmental impact reduction. But there are a variety of O&M programs which have been developed for existing building owners, and although not specifically referred to as "high-performance," they have high-performance characteristics and we will review some of the principal ones here.

General Services Administration (GSA)

The GSA has a very useful and straightforward program called the Sustainable Facilities Tool (SFT) which, although not a compliance path or certified procedure, includes a section on sustainable building operations and maintenance. It is a public-access, no-cost tool, and it includes a general checklist which the building owner or manager can use to profile the building and adopt strategies for high-performance results.[3]

Sustainable Strategies—Sustainability In Action

Education is the key to changing behavior: provide training for building operators and occupants and help them support sustainable workplace initiatives.

Provide information regarding what has been done and how they can support the following sustainable efforts:

- Indoor air quality (adequate outside air, high quality filters)
- Resource conservation (water, energy, materials)
- Waste reduction (recycling)
- Responsible procurement (ecologically sensitive materials/vendors)
- Transportation (mass transit and alternative commuting)

Operations Best Practices
- Seek opportunities that leverage the scale and diversity of buildings as well as the depth and breadth of experience and operational expertise across the federal government
- Test promising sustainable technologies in real-life circumstances to implement innovative products, services and ideas that offer the broadest possible value to the federal government, industry, communities and environment
- Provide quick response to occupants' comfort concerns to help with occupant satisfaction
- Provide feedback to occupants on energy use to help adjust their behaviors to help achieve energy goals

Indoor Air Quality
- Ensure the building maintenance staff is conducting the preventive maintenance on all building exhaust systems (restrooms, garage exhaust fans, etc).
- Adjust occupancy based systems like HVAC and ventilation to actual occupancy times of the building
- Retrofit existing cooling systems to run on environmentally friendly HFCs

Energy Conservation
- Monitor energy consumption by using the ENERGY STAR Portfolio Manager; with precise thresholds for HVAC operations to ensure energy efficiency, comfort, operations and productivity
- Monitor unnecessary electrical demand and ensure it is curtailed immediately
- Adjust occupancy based systems like HVAC and ventilation to actual occupancy times of the building
- Provide high degree of responsiveness for equipment malfunctions. Ask for occupant cooperation in turning off lights as employees leave in the evening and on weekends. At the same time encourage cleaning staff to turn off the lights in these same areas once cleaning is completed
- Explore using day time cleaning to improve air quality and cleanliness of office spaces
- Install motion sensors in utility closets, mechanical rooms and restrooms when space is not used
- Set guidelines for high impact energy use days in summer or winter months with practices such as closing or lowering blinds, turning off equipment and lights that are not in use all the time or when vacant such as conference, copy and kitchens
- Schedule large copier jobs for off-peak times, usually early morning
- Replace old fluorescent lighting with, for example, high-efficiency low-mercury T8/T5 and compact fluorescent bulbs

Water Management
- Consider sub-metering of water systems to identify future opportunities for improvement
- Provide training for occupants use of water efficient fixtures
- Provide training for proper maintenance of waterless urinals

Water Reduction and Management
- Use electronic communications versus printed material for facilities guidelines and announcements
- Institute a program to recycle materials and investigate use of mixed or

co-mingled paper, newspapers, aluminum cans and plastic bottles
- Hold trash and recycling workshops to education occupants about what can be recycled and how it's collected
- Provide desk-side and kitchen/copier room recycling containers and extra containers for trash and recycling for major office clears
- Integrate waste reduction and management with local municipalities
- Divert construction waste from landfills and facilitate the recycling of construction waste through the standardization of processes

Custodial Services
- Require cleaning contractors to implement green cleaning practices including and not limited to using resources for recycled content paper stock and trash liners, environmentally friendly cleaning products, and vacuum cleaners with filtration systems
- Use cleaning products that meet Green Seal Standard GS-37, products with low volatile organic compound (VOC) levels and products with high post-consumer recycled content
- Conduct cleaning audits to ensure cleaning is done properly
- Use less or non-toxic chemicals or remove pest harborages to manage pests
- Explore using day time cleaning to improve air quality and cleanliness of office spaces

The tool expands to give more detailed information and links on specific topics, for example:

Flush Fixtures[4]

Toilets and urinals contribute to approximately one third of the total building water consumption. Water efficient fixtures such as ultra-low flow, dual flush, pressure-assisted, or waterless technologies can lead to significant savings. Conventional toilets and urinals require 1.6 or 1.0 gallons per flush (gpf) respectfully, while their efficient counterparts use little (0.125 – 1.0 gpf) to no water at all. Utilizing non-potable water sources, such as gray water, for flushing further diminishes overall potable water consumption. Maintenance of flush valves is imperative, as a constantly running toilet can waste over 200 gallons of water each day (EPA WaterSense: The Facts on Leaks).

Or, another example:

Single-Pass Cooling Equipment[5]

Single-pass (or once-through) cooling equipment uses water for only one cycle before subsequently discharging it. Examples of this equipment include condensers, air compressors, degreasers, vacuum pumps, ice machines, and air conditioners. In order to maximize water savings, single-pass systems should be eliminated altogether or be replaced or modified to operate on a closed-loop that recirculates water instead of discharging it. If a closed-loop alternative is not feasible, implementing an automatic shut-off valve and reusing the discharged water for non-potable situations are sustainable alternatives. To remove the same heat load, single-pass systems use 40 times more water than a cooling tower operated at five cycles of concentration.* To maximize water savings, single-pass cooling equipment should be either modified to recirculate water or, if possible, should be eliminated altogether. See FEMP BMP 9.

*Federal Energy Management Program (FEMP) BMP 9: Single-Pass Cooling Equipment

---

Federal Energy Management Program (FEMP)

The FEMP, Release 3.0, *Operations & Maintenance Best Practices: A Guide to Achieving Operational Efficiency*, published in August 2010[6], is a public access, no-cost, 321-page document, which outlines O&M procedures for federal buildings. Some of the procedures are based on earlier work commissioned by the Department of Defense in the 1990s. These are not specifically referred to as high-performance measures, but the prescriptions are all in the same vein.

The FEMP O&M Guide (FOMG) is based on the fundamental concept of OMETA[7], O&M Integration, and includes operation, maintenance, engineering, training, and administration.

The FOMG outlines the following:

- O&M management issues and their importance
- Computerized maintenance management systems (CMMS)
- Types of maintenance programs and definitions
- Maintenance technologies and accepted predictive technologies
- Building commissioning processes and O&M
- Metering
- O&M procedures for equipment at most federal facilities and calculation procedures for estimating energy savings.
- New O&M technologies
- Ten steps to initiating an operational efficiency program

The FOMG is in direct response to the passage of E.O. 13514, *Federal Leadership in Environmental, Energy & Economic Performance*[8] signed into law October of 2009. The order directed federal agencies to further address energy, water, and operational efficiency with key targets as follows[9]:

- Federal agencies must enhance efforts toward sustainable buildings and communities. Specific requirements include the implementation of *high-performance sustainable* federal building design, construction, *operation and management*, maintenance, and deconstruction.

- Pursuing cost-effective, innovative strategies (e.g., highly reflective and vegetated roofs) to minimize consumption of energy, water, and materials.

- Managing existing building systems to reduce the consumption of energy, water, and materials, and identifying alternatives to renovation that reduce existing asset-deferred maintenance costs.

- Reducing potable water consumption intensity 2% annually through FY 2020, or 26% by the end of FY 2020, relative to a FY 2007 baseline.

- Reducing agency industrial, landscaping, and agricultural water consumption 2% annually, or 20% by the end of FY 2020, relative to a FY 2010 baseline.

- Identifying, promoting, and implementing water reuse strategies consistent with state law that reduce potable water consumption.

In discussing the tremendous potential for O&M programs to produce energy and cost savings with minimal cash outlays, the FOMG sources two important studies:

Energy Savings Expert Teams[10]
- To realize the same benefits (energy savings), equipment retrofits cost approximately 20-times more than low-cost O&M measures
- Dollars saved per dollars invested (calculated values):
  — O&M projects: 3.83 (simple payback 0.26 years)
  — Retrofit projects: 0.19 (simple payback 5.26 years)

- Overall program cost-effectiveness for measures implemented (as of May 2007) includes retrofit, O&M measures and program administration/delivery
  — Annual energy savings: 202,512 MMBtu
  — Annual cost savings: $1,731,780
  — Total program cost: $1,795,000
  — **Simple payback: 1.0 years**

Energy Efficiency Expert Evaluation—E4[11]
- Calculated savings range from 3% to over 40%, average savings 15%.
- Dollars saved per dollars invested (calculated values):
  — O&M projects (defined as <$5,000): 14.9 (simple payback 0.07 years)
  — Retrofit projects (defined as >$5,000): 0.7 (simple payback 1.5 years)

- Overall program cost effectiveness for measures implemented (as of December 2008), includes retrofit, O&M measures and program administration/delivery
  — Annual cost savings: $584,000
  — Total program cost: $800,000
  — **Simple payback: 1.4 years**

The following is presented as a specific case study[12]:

A demonstration focused on O&M-based energy efficiency was conducted at the U.S. Department of Energy Forrestal Building in Washington, D.C. (Claridge and Haberl 1994). A significant component to this demonstration was metering and the tracking of steam use in the building. Within several months, $250,000 per year in steam leaks were found and corrected. These included leaks in a steam converter and steam traps. Because the building was not metered for steam and there was not a proactive O&M program, these leaks were not detected earlier, nor would they have been detected without the demonstration. The key lessons learned from this case study were:

- O&M opportunities in large buildings do not have to involve complex engineering analysis.
- Many O&M opportunities exist because building operators may not have proper documentation that hindered day-to-day actions.
- Involvement and commitment by building administrators is a key ingredient for a successful O&M program.

As a building owner or manager, simple paybacks of 1.0 and 1.4 years should get your attention. These are more examples of how waste reduction (in this case stopping leakage) is Job #1 and how good O&M procedures can help meet that goal.

Leadership in Energy and Environmental Design—
Existing Buildings, Operations and Maintenance (LEED-EB O&M)

LEED-EB O&M[13] is a fee-based voluntary rating system promulgated by the United States Green Building Council for assessing and promoting sustainable high-performance measures in existing buildings. It has been available for adoption since 2009 and was revised most recently in July, 2012.

The LEED-EB O&M rating system is one of the most comprehensive approaches to O&M, and as with most LEED certifications, it requires a large amount of documentation for certification and validation. But the system has been seeing wider implementation, especially when major renovations to existing buildings are engaged, and there are numerous opportunities to bring the design-oriented and environmental impact approach which LEED favors. In addition to this, numerous municipalities have already adopted LEED standards (or recognized equivalent) into their green building ordinances, so using LEED-EB O&M may be a logical next step for them. The GSA, for example, has recently raised its standard for ALL new federal buildings to LEED Gold.[14] What this means is that LEED is gaining market traction both with federal and municipal government agencies as well as its growing body of Class A commercial building applications. The checklist shown below outlines the basic prescriptions of the system.[15]

LEED 2009 for Existing Buildings: Operations and Maintenance Project Checklist

Sustainable Sites 26 Possible Points

☐ Credit 1 LEED Certified Design and Construction 4
☐ Credit 2 Building Exterior and Hardscape Management Plan 1
☐ Credit 3 Integrated Pest Management, Erosion Control, and Landscape Management Plan 1
☐ Credit 4 Alternative Commuting Transportation 3-15
☐ Credit 5 Site Development—Protect or Restore Open Habitat 1
☐ Credit 6 Stormwater Quantity Control 1
☐ Credit 7.1 Heat Island Reduction—Nonroof 1
☐ Credit 7.2 Heat Island Reduction—Roof 1
☐ Credit 8 Light Pollution Reduction 1

Water Efficiency 14 Possible Points

☑ Prerequisite 1 Minimum Indoor Plumbing Fixture and Fitting Efficiency Required
☐ Credit 1 Water Performance Measurement 1-2
☐ Credit 2 Additional Indoor Plumbing Fixture and Fitting Efficiency 1-5
☐ Credit 3 Water Efficient Landscaping 1-5
☐ Credit 4.1 Cooling Tower Water Management—Chemical Management 1
☐ Credit 4.2 Cooling Tower Water Management—Nonpotable Water Source Use 1

(Continued)

Checkist (*concluded*)

Energy and Atmosphere 35 Possible Points

☑ *Prerequisite 1 Energy Efficiency Best Management Practices—Planning, Documentation, and Opportunity Assessment R equired*

☑ *Prerequisite 2 Minimum Energy Efficiency Performance Required*

☑ *Prerequisite 3 Fundamental Refrigerant Management Required*

☐ *Credit 1 Optimize Energy Efficiency Performance 1-18*

☐ *Credit 2.1 Existing Building Commissioning—Investigation and Analysis 2*

☐ *Credit 2.2 Existing Building Commissioning—Implementation 2*

☐ *Credit 2.3 Existing Building Commissioning—Ongoing Commissioning 2*

☐ *Credit 3.1 Performance Measurement—Building Automation System 1*

☐ *Credit 3.2 Performance Measurement—System Level Metering 1-2*

☐ *Credit 4 On-site and Off-site Renewable Energy 1-6*

☐ *Credit 5 Enhanced Refrigerant Management 1*

☐ *Credit 6 Emissions Reduction Reporting 1*

Materials and Resources 10 Possible Points

☑ *Prerequisite 1 Sustainable Purchasing Policy Required*

☑ *Prerequisite 2 Solid Waste Management Policy Required*

☐ *Credit 1 Sustainable Purchasing—Ongoing Consumables 1*

☐ *Credit 2.1 Sustainable Purchasing—Electric-Powered Equipment 1*

☐ *Credit 2.2 Sustainable Purchasing—Furniture 1*

☐ *Credit 3 Sustainable Purchasing—Facility Alterations and Additions 1*

☐ *Credit 4 Sustainable Purchasing—Reduced Mercury in Lamps 1*

☐ *Credit 5 Sustainable Purchasing—Food 1*

☐ *Credit 6 Solid Waste Management—Waste Stream Audit 1*

☐ *Credit 7 Solid Waste Management—Ongoing Consumables 1*

☐ *Credit 8 Solid Waste Management—Durable Goods 1*

☐ *Credit 9 Solid Waste Management—Facility Alterations and Additions 1*

Indoor Environmental Quality 15 Possible Points

☑ *Prerequisite 1 Minimum Indoor Air Quality Performance Required*

☑ *Prerequisite 2 Environmental Tobacco Smoke (ETS) Control Required*

☑ *Prerequisite 3 Green Cleaning Policy Required*

☐ *Credit 1.1 Indoor Air Quality Best Management Practices—Indoor Air Quality Management Program 1*

☐ *Credit 1.2 Indoor Air Quality Best Management Practices—Outdoor Air Delivery Monitoring 1*

☐ *Credit 1.3 Indoor Air Quality Best Management Practices—Increased Ventilation 1*

☐ *Credit 1.4 Indoor Air Quality Best Management Practices—Reduce Particulates in Air Distribution 1*

☐ *Credit 1.5 Indoor Air Quality Best Management Practices—Indoor Air Quality Management for Facility Alterations and Additions 1*

☐ *Credit 2.1 Occupant Comfort—Occupant Survey 1*

☐ *Credit 2.2 Controllability of Systems—Lighting 1*

☐ *Credit 2.3 Occupant Comfort—Thermal Comfort Monitoring 1*

☐ *Credit 2.4 Daylight and Views 1*

☐ *Credit 3.1 Green Cleaning—High Performance Cleaning Program 1*

☐ *Credit 3.2 Green Cleaning—Custodial Effectiveness Assessment 1*

☐ *Credit 3.3 Green Cleaning—Purchase of Sustainable Cleaning Products and Materials 1*

☐ *Credit 3.4 Green Cleaning—Sustainable Cleaning Equipment 1*

☐ *Credit 3.5 Green Cleaning—Indoor Chemical and Pollutant Source Control 1*

☐ *Credit 3.6 Green Cleaning—Indoor Integrated Pest Management 1*

Innovation in Operations 6 Possible Points

☐ *Credit 1 I nnovation in Operations 1-4*

☐ *Credit 2 LEED Accredited Professional 1*

☐ *Credit 3 Documenting Sustainable Building Cost Impacts 1*

Regional Priority 4 Possible Points

☐ *Credit 1 R egional Priority 1-4*

Like all LEED rating systems, LEED-EB O&M has a scoring system and a minimum number of points for accreditation at the different levels of certification. There is also a significant and obvious emphasis on environmental and human impact with regard to purchasing of materials, IAQ and green cleaning. There is a big emphasis on overall indoor environmental quality and the attendant benefits for the occupants. This can be seen in the total number of points available for sustainable sites, materials and resources and indoor environmental quality: 41 points, vs. 35 points for energy and atmosphere.

LEED-EB O&M does have some significant restrictions to its application as evidenced here in Section 5[16]:

V. Facility Alterations and Additions

Although LEED for Existing Buildings: Operations and Maintenance focuses mainly on sustainable ongoing building operations, it also embraces sustainable alterations and new additions to existing buildings.

In general parlance, alterations and additions may range from a complete gutting, major renovation, or large new wing to the replacement of an old window, sheet of drywall, or section of carpet.

In LEED for Existing Buildings: O&M, however, alterations and additions has a specific meaning. It refers to changes that affect usable space in the building. Mechanical, electrical, or plumbing system upgrades that involve no disruption to usable space are excluded.

Only alterations and additions within the following limits are eligible for inclusion in LEED for Existing Buildings: O&M certification:

- Maximum. Alterations that affect no more than 50% of the total building floor area or cause relocation of no more than 50% of regular building occupants are eligible. Additions that increase the total building floor area by no more than 50% are eligible. Buildings with alterations or additions exceeding these limits should pursue certification under the LEED for New Construction program.

- Minimum. Alterations that include construction activity by more than 1 trade specialty, make substantial changes to at least 1 entire room in the building, and require isolation of the work site from regular building occupants for the duration of construction are eligible. Additions that increase the total building floor area by at least 5% are eligible. Alterations or additions below these limits are considered repairs, routine replacements, or minor upgrades and are ineligible to earn points under LEED for Existing Buildings: O&M.

The minimum applies to Materials and Resources (MR) Credits 3 and 9, and Indoor Environmental Quality.

The conclusion is that the LEED-EB O&M certification is designed for more holistic, integrated applications, and is not suited for upgrades to mechanical and electrical equipment only as part of O&M procedures.

MONITORING

In the new construction or retrofit of a HP building, we are first concerned with using the design and engineering process to minimize resource flows, costs and environmental impact in the process. In existing buildings we are concerned with monitoring resource flows, costs and environmental impacts to track on-going operations and insure high-performance results in the existing or potentially retrofitted infrastructure. Continuous monitoring is what gives us the regular feedback needed to evaluate and improve, not only the operations of the existing building, but also future design and improvements. Continuous improvement then occurs using an O&M schema like the one we developed earlier in this chapter, coupled with monitoring equipment, reporting systems and education. All of the codes, standards and rating systems concerned with "high performance" we reviewed in Chapter 3 require monitoring and reporting.

ENERGY

There are literally hundreds of monitoring systems, or building energy management systems (BEMS), on the market. These are all computer controlled and employ remote sensing devices to feed information to a central data processing unit. Many of them now are integrating the monitoring of energy, water and waste data points into a central database. The overall goal is to "sensitize" all building functions that can then provide continuous, real-time feedback on operations. Or as Dr. Richard Watson of the University of Georgia has said, referring to the emerging field of Energy Informatics[17]: "…we need to sensitize all of the assets in the (building) environment." This follows the idea that a HP building is more like a "living system," which can respond instantaneously to the environment and communicate with the users.

Although there are hundreds of systems available, they are structured similarly in the way they integrate building functions with IT networks, as in Figure 10-3 from Talisen,[18] or Figure 10-4 from Spara.

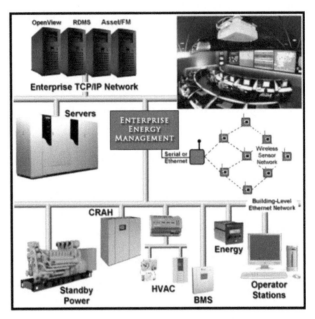

Figure 10-3.

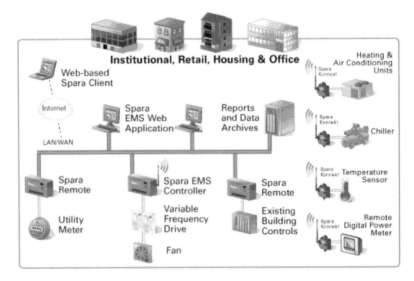

Figure 10-4.

To use BEMS, it is assumed that the building is metered, sub-metered and set up with data-loggers. The information can then be collated into a centralized system on-site, or shipped to on-line services which provide software platforms, dashboards and cloud storage. Figure 10-5 gives a dashboard example from Lucid Technologies.[50]

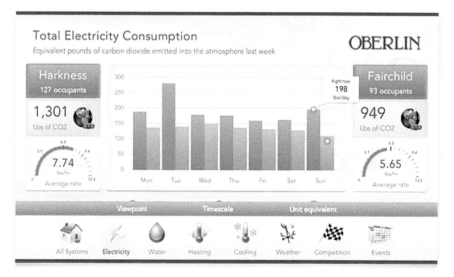

Figure 10-5.

Consider the following from Pike Research, as quoted in Business Wire[21]:

Building Energy Management Systems Investment to Top $10 Billion in the United States by 2016, Forecasts Pike Research
BOULDER, Colo.—(BUSINESS WIRE)—Commercial and residential buildings are the largest consumers of electricity in the United States, representing nearly two-thirds of all electricity used. As property owners and managers look for ways to reduce their energy expenditures and environmental impact through efficiency measures, they are increasingly adopting "smart building" technologies including automation and controls, high-efficiency systems and equipment, and a broadening array of energy management services. These enhancements are being unified in Building Energy Management Systems (BEMS), which are information technology (IT)-based hardware and software systems being employed in a holistic approach to optimize the efficiency of buildings' energy consumption. According to a recent report from Pike Research, the BEMS market holds significant opportunity for growth in the next few years, and the firm forecasts

that investment in the sector will total $10.1 billion in the U.S. during the period between 2010 and 2016, with a compound annual growth rate (CAGR) of 17.4%.

"When it comes to energy use, commercial buildings are getting smarter all the time," says research analyst Jevan Fox. "This intelligence, which encompasses everything from sensor networks to predictive supply and demand algorithms to high-efficiency HVAC systems, will require a greater level of control and coordination than legacy building management systems (BMS) can provide. The benefits of BEMS will create a growing market opportunity for a variety of industry players, ranging from traditional building systems companies and energy service companies (ESCOs) to the expanded influence of IT vendors on facilities management."

WATER

The only thing more important than energy to the functioning of an HP building is water. And it is worth pointing out the following. Unlike the electrical utility which only charges you for electrical energy when it enters the building and not for the waste heat that is rejected or leaves the building, you are charged by the water utility whenever water enters the building *and whenever it leaves and enters the municipal wastewater system*. You are also paying for potable water to run toilets and irrigation systems when you don't need that kind of water quality for those functions. There is significant demand pressure increasing on water supplies and on water utilities to provide enough water for buildings and, in certain regions, for industrial processes like natural gas hydraulic fracturing (which uses huge volumes of water). This demand pressure is only going to increase and with it the price for water. This is all the more reason that HP buildings have to economize water usage by:

[1]   Using low-flow plumbing fixtures[22], high-efficiency water chillers, drip irrigation systems[23] and on-site drinking water purification[24]

[2]   Economizing flow rates with metering and operator and occupant education

[3]   On-site collection and recycling of municipal water, process water, and rainwater

Ultimately gray water collection and reuse is fundamental, since gray water, with minimal processing, can be used to flush toilets and irrigate landscaping, and with some additional processing could be used to run chillers or boilers. It is also pointless to make the municipal water utility process gray water in a wastewater treatment facility when it can be re-used at the building. The same thing is true for rainwater, which can be collected, stored and re-used rather than letting it run off onto adjacent properties or go into the storm sewer. Jurisdictions in the US which have adopted the International Plumbing Code (IPC) are eligible to reuse gray water for irrigation purposes and for flushing toilets. The principles of these types of systems are similar, such as the rainwater collection system in Figure 10-6[25] or the gray water system[26] in Figure 10-7.

Planning for rainwater collection and gray water recycling in a new building is relatively straightforward and is increasingly part of the design of HP buildings (a lot of these applications occurring in residential construction where water usage density is lower). Obviously it

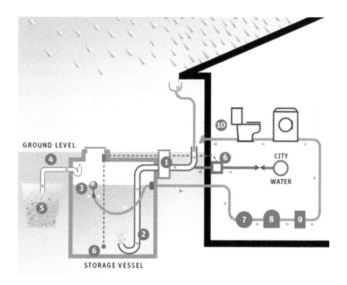

Figure 10-6.

1: RainKeeper self-cleaning filter.	6: Integration controller.
2: Calmed inlet.	7: Electronic pump control.
3: Floating intake.	8: Pump.
4: Overflow Siphon.	9: Pressure tank.
5: Infiltration.	10: Rainwater for washer, toilets, lawn

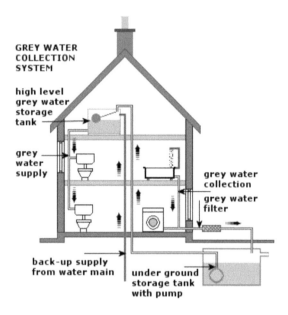

Figure 7.

may be no simple task to retrofit an existing building with storage tanks and piping that re-circulates stored and filtered or treated water. For this reason, as with distributed energy generation, there is movement towards the idea of *high-volume water filtering or treatment* on-site, or *distributed wastewater infrastructure*. There is a growing installed base of these types of systems in industrial settings, which could in principle be converted for use in commercial applications. The market for these systems in residential and commercial buildings is near the beginning of the curve but is showing strong growth potential with the increase in demand and pricing pressures. There are numerous systems available and we consider a few below.

Rotary Disk

Rotary disk filter systems can be designed to remove almost any type of particulate matter down to 1 micron in diameter. The principles are very similar from one product to another, as seen in Figures 10-8 and 10-9. An actual system in place from Kadant Petax for paper manufacturing which can process at 1056 gpm (gallons per minute), as shown by Figure 10-10.

SCHEMATIC OF DISC FILTER

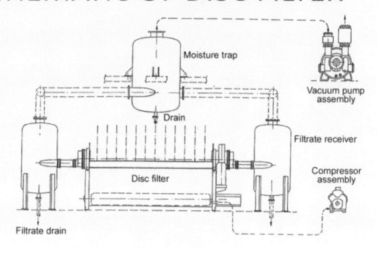

Figure 10-8.[27]

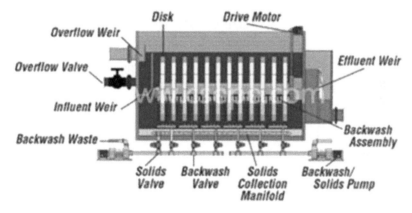

Figure 10-9.[28]

Membrane

Membranes as a technology for wastewater treatment are emerging as an important player in the market for on-site processing and can be implemented down to .1 - .2 micron pore size with MF membranes and even .005 - .01 with some UF membranes.[29] Significant advances in the last ten years have positioned the technology to advance rapidly

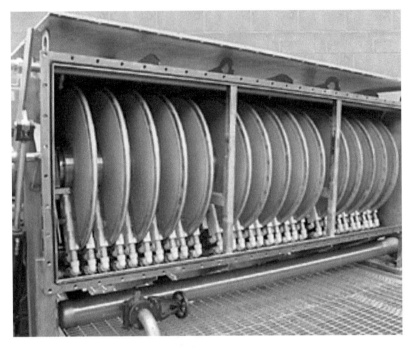

Figure 10-10.

due to their efficient throughput and space-saving advantages. Membrane technologies can include the following:

- Membrane bioreactors—usually microfiltration (MF) or ultrafiltration (UF) membranes immersed in aeration tanks (vacuum system), or implemented in external pressure-driven membrane units, as a replacement for secondary clarifiers and tertiary polishing filters.

- Low-pressure membranes—usually MF or UF membranes, either as a pressure system or an immersed system, providing a higher degree of suspended solids removal following secondary clarification. UF membranes are effective for virus removal.

- High-pressure membranes—nanofiltration or reverse osmosis pressure systems for treatment and production of high-quality product water suitable for indirect potable reuse and high-purity industrial process water. Also, recent research has shown that microconstituents, such as pharmaceuticals and personal care products, can be removed by high-pressure membranes.[30]

Figure 10-11 is a schematic from ProChem, and on-site application for natural gas production[31] is shown in Figure 10-12.

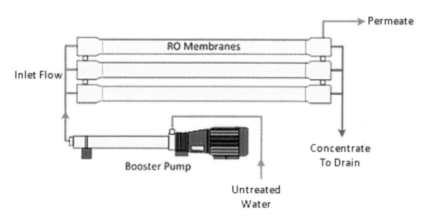

Figure 10-11

Figure 10-12

Shared Services

Assessing flow rates is, of course, critical in designing any water treatment systems for commercial applications, since their continuous high-volume demand schedules and water quality requirements may be quantitatively and characteristically different than industrial process environments. But this should not provide any serious obstacles. As with energy, increasing demand and cost will ultimately stimulate technological development and economies of scale for commercial system production. Additionally, buildings in close proximity to one another could leverage scalar advantages by combining water treatment process requirements in a local cluster. This approach has been taken in some commercial settings and on some military bases where there are a number of buildings in the same area which could take advantage of "regional" distributed wastewater processing.

Figure 10-13 gives a commercial example from Aquapoint.[32]

MATERIALS

Resource flows of paper, plastics and chemicals into a building are also a major area of importance for high-performance O&M. Behavioral

Figure 10-13
This Aquapoint Bioclere™ wastewater treatment plant in a suburb south of Boston serves three commercial shopping plazas and an office complex.

changes can have a huge impact on consumption patterns before engineering changes are implemented for equipment upgrades or process improvements. "Green purchasing" or "sustainable procurement" involves the purchasing of materials which are recycled, non-toxic, reduce environmental impact and which can strategically reduce flows in, and waste out of, the building. Changing single-use materials into multiple-use materials wherever possible is also essential. All of that is *saving you money*.

A simple way to look at controlling materials flow in, and thus seriously reducing waste out and the necessity for recycling is with the "waste hierarchy pyramid" shown in Figure 10-14.[33]

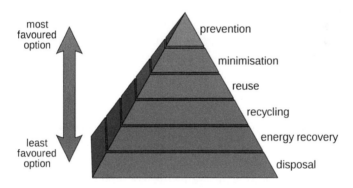

Figure 10-14

In other words, if you cut the requirement for the material in the first place, then you don't need to handle it, process it, recycle it or dispose of it. Recycling is a good thing but it is 4th from the top in terms of priority.

Product Usage

Many of the products we use in a building are single-use but could be converted to multiple-use quite easily. These are very simple, common-sense measures but it is amazing how often we do not implement them because of habit or convenience. There are dozens of examples and we list just a few here:

- reuse beverage containers with low recyclability such as paper cups

- reuse plastic cups for beverages or plastic containers for cleaning products (refill cleaning product containers used by service personnel from centralized filling stations)

- print all documents for internal use on two-sided paper and/or reuse paper already printed on one side. Use recycled paper for all internal use and virgin paper only for presentation documents.

- filter drinking water at drinking fountains and/or install stand-alone water purification stations which use municipal water

Consider the following. It is estimated that in the US we used approximately 23 billion paper cups in 2010.[34] Since paper cups are almost exclusively single-use and are difficult to recycle, nearly all of them end up as waste in the landfill. But do you really need to throw the paper cup away after you use it once? Almost all paper cups can be used twice without the cup degrading. If you could get half the people, half the time, to use paper cups twice, you would remove approximately 6 billion cups from the waste stream. This approach works with other supposedly single-use products as well: plastic cups, plastic cleaning products containers, etc.

The Container Recycling Institute[35] publishes important research and statistics on container usage in the US which can inform the conduct of O&M in a HP building. Figures 10-15 and 10-16 let us look at type, overall usage, per capita usage and wasting.

The data show quite clearly that we have a "container problem" in the US. Since all beverage containers, aluminum, glass, plastic, steel, whether recycled or not, are single use, and per capita usage is increasing, measures must be instituted to reduce the resource flows associated with this avalanche of waste. Bottled water is a particularly egregious example of wasted resources, as shown in Figure 10-17.

There is no reason, in principle, to import bottled water into a facility, when municipal water can easily and cost-effectively be filtered on-site and drunk from re-usable containers.

Programs

There are two organizations besides the GSA and the FEMP which stand out when it comes to programs for managing material flows, the Responsible Purchasing Network (RPN)[36] and the Global

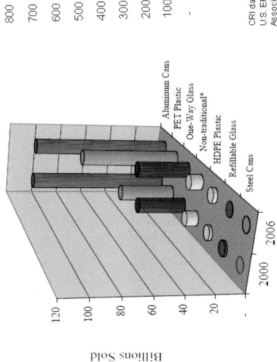

Estimated U.S. Per Capita Consumption of Beverage Containers, 1970-2005

CRI data derived from Aluminum Association, U.S. Commerce Dept, U.S. EPA Office of Solid Waste, American Plastics Council, National Association of PET Container Resources. Includes aluminum, steel, glass, PET plastic, HDPE plastic. Includes dairy.

© Container Recycling Institute, 2006

Sales by Container Type, 2000 and 2006

* Non-Traditional containers include aseptic boxes, gable-top cartons, and foil pouches.
© Container Recycling Institute, 2008.

Figure 10-15

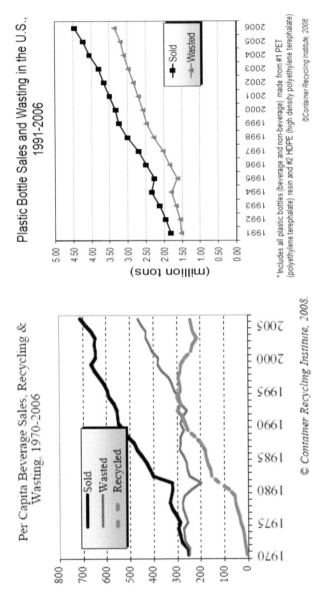

Figure 10-16

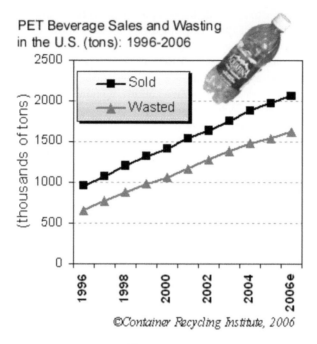

Figure 10-17

Environmental Management Initiative (GEMI)[37]. Both of these are widely subscribed to and have as members[38] some of the major materials purchasers[39] in industry and business. Both organizations produce ongoing research and multiple tools which can be used by Owners and Managers to track and analyze purchasing and materials management decisions.

RPN produces purchasing guides for a wide range of products and services, including: bottled water, carbon offsets, cleaners, computers, copy paper, fleets, food services, green power, light-duty tires and wheel weights, lighting, office electronics, paint, and toner cartridges. Some text examples from the guide on bottled water and the guide on toner cartridges follow.

Americans bought a total of 8.8 billion gallons of bottled water in 2007. According to one estimate, producing these bottles required the energy equivalent of over 17 million barrels of oil and produced over 2.5 million tons of carbon dioxide. This is the same amount of carbon dioxide that would be emitted by over 400,000 passenger vehicles in one year. Nearly 50 billion new PET (polyethylene terephthalate) plastic bottles were produced in 2005 from virgin rather than recycled materials, producing additional greenhouse gases. In 2004, only 14.5 percent of non-carbonated beverage bottles made from PET were recycled. For each gallon of water that is bottled, an additional two gallons of water are used in processing. Many of these impacts can be easily avoided by switching to tap water, filters, fountains and coolers when necessary.

Each year, over 350 million toner cartridges go to landfills and incinerators in the United States. A typical OEM toner cartridge consumes 5 to 9 pounds of virgin material in the production process and is composed of 40% plastic and 40% metal. Remans (remanufacturers) reuse this plastic and metal and save other materials and energy needed to produce a comparable OEM cartridge. Cartridge remanufacturers in the United States reuse over 35 thousand tons of plastic and save over 400,000 barrels of oil each year. Since cartridges may be remanufactured more than once, resource intensity is reduced further with each additional remanufacture by extending the total life of the unit.

GEMI publishes numerous tools including the GEMI Solution Tool Matrix[40], a broad-based approach that allows users to examine multiple areas for assessment and do long-range planning. They also produce on-line interactive tools for a Water Sustainability Planner, a Supply Chain Tool[41], and a Health, Safety and Environment Management Information Systems (HSE-MIS).[42]

SUMMARY

Controlling resource flows in an out of the HP building requires detailed, systematic O&M planning, sensitizing building assets, metering, monitoring, data analytics and continuous feedback to owners and facilities managers. This integrated systems approach leads to continuous operational efficiency and improvement. It also allows benchmarking against other similar building types and a community which learns from the dynamic environment of resource delivery systems, building operation, occupant behavior and waste stream processing.

References

1. http://en.wikipedia.org/wiki/Ludwig_Mies_van_der_Rohe
2. http://sftool.gov/plan/400/life-cycle-assessment-lca-overview
3. http://sftool.gov/plan/268/sustainable-building-operations-and-maintenance-services
4. http://sftool.gov/explore/green-building/section/40/water/resources-impact
5. http://sftool.gov/explore/green-building/section/40/water/resources-impact
6. http://www1.eere.energy.gov/femp/pdfs/omguide_complete.pdf
7. SEE: Meador, R.J. 1995. Maintaining the Solution to Operations and Maintenance Efficiency Improvement. World Energy Engineering Congress, Atlanta, Georgia.
8. http://www.whitehouse.gov/assets/documents/2009fedleader_eo_rel.pdf
9. O&M Best Practices Guide, Release 3.0, p. 23.
10. See: Hunt, D. 2007. Energy Savings Expert Team (ESET) Benefits Assessment. Presentation to the Federal Energy Management Program, May 2007. Pacific Northwest National Laboratory, Richland, Washington.
11. SEE: Benefits of the 2007 Energy Efficiency Expert Evaluations (E4). Hail, J., Presentation to the Federal Energy Management Program, December 2009. Pacific Northwest National Laboratory, Richland, Washington.
12. O&M Best Practices Guide, Release 3.0, p. 27.
13. http://new.usgbc.org/sites/default/files/LEED%202009%20Rating_EBOM-GLOBAL_07-2012_8d_0.pdf
14. http://www.gsa.gov/portal/content/105251
15. LEED 2009 Rating_EBOM Global_7-2012_8d_0.pdf, p. 8-9
16. LEED 2009 Rating_EBOM Global_7-2012_8d_0.pdf, p. 20.
17. http://energyinformatics.info/
18. http://www.talisentech.com/
19. http://www.poweritsolutions.com/commercial/
20. http://www.luciddesigngroup.com/products.php
21. http://www.businesswire.com/news/home/20110408005011/en/Building-Energy-Management-Systems-Investment-Top-10
22. Extensive information on water conservation with low-flow fixtures is available from: http://water.epa.gov/polwaste/nps/chap3.cfm, and on good planning for adopting low-flow fixtures: http://www.facilitiesnet.com/plumbingrestrooms/article/Water-Conservation-Use-Utility-Bills-to-Establish-Baseline—11936
23. For an excellent and very comprehensive guide to drip irrigation for landscaping see: http://www.amwua.org/pdfs/drip_irrigation_guide.pdf, published by the Arizona Municipal Water Users Association: http://www.amwua.org/
24. For extensive information regarding on-site water purification techniques and standards, see: http://www.nsf.org/
25. http://www.starkenvironmental.com/images/cistern_diagram.jpg
26. http://www.desalresponsegroup.org/images/grey-water-storage.gif
27. http://www.headingfilter.com/
28. http://en.csepe.com/products_detail/&productId=c999162e-9b27-4100-beb0-c5e320fbc183.html
29. See: American Membrane Technology Association - http://www.amtaorg.com/
30. http://cdmsmith.com/en-US/Insights/Viewpoints/Membrane-Technology-Advances-Wastewater-Treatment-and-Water-Reuse.aspx
31. http://www.prochemweb.com/systemsGallery.html
32. http://www.aquapoint.com/images/LOWES_PEMBROKE_3-5-08.pdf
33. http://commons.wikimedia.org/wiki/File:Waste_hierarchy.svg

34. http://www.greendesignetc.net/GreenProducts_10/GreenProducts_Park_Jong-pil_Paper.pdf
35. http://www.container-recycling.org/index.php/factsstatistics/allcontainers
36. http://www.responsiblepurchasing.org/about/index.php
37. http://www.gemi.org/gemihome.aspx
38. http://www.responsiblepurchasing.org/about/partners.php
39. http://www.gemi.org/GEMIMemberCompanies.aspx
40. http://www.gemi.org/toolmatrix/
41. http://www.gemi.org/supplychain/resources/ForgingNewLinks.pdf
42. http://www.gemi.org/GEMIInteractiveTools.aspx

Chapter 11

Outputs

High-performance is an integrated systems approach to the design, engineering, construction, operation and maintenance of buildings. In general it requires:

[1]   The assessment of material properties and performance characteristics from all areas of the building

[2]   The consideration of building impact on the environment generally and the external microclimate specifically

[3]   The systems engineering and continuous analysis of resource flows, information processing, occupant behavior and facilities management

[4]   A strategy of diverse energy resource inputs

[5]   A monitoring system that creates an expanding knowledge economy about building function and performance

We know that the design, engineering, construction and operation of high performance buildings can benefit from the techniques of concurrent engineering in manufacturing which emphasize a cyclical approach with continuous feedback loops. A building is not a manufactured product, but it has a measurable life-cycle just the same. See Figure 11-1.

High-performance buildings increase resource efficiency, improve cost efficiency and reduce environmental impact by:

[1]   decreasing resource inputs upstream by cutting waste, re-using resources on-site and reducing demand through proper design, engineering and management of the building

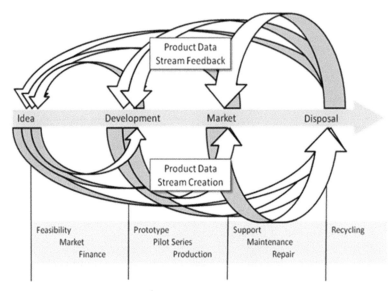

Figure 11-1[1]

[2]   establishing process improvements in operations through education and continuous commissioning

[3]   making capital investments in envelope modifications and mechanical and electrical equipment

[4]   diversifying energy inputs by changing demand curves, collecting energy on-site and utilizing distributed energy generation (whether renewable or fossil-fuel based)

[5]   having net-zero energy and zero-waste as overall goals

These techniques help building owners and managers continuously eliminate waste, increase efficiencies, improve costs and reduce environmental impact. In certain contexts, high-performance buildings may also become net-energy generators, producing more energy than they consume and selling it back to the grid or distributing it locally. We have seen that energy efficiency and environmental impact are of primary concern, but that waste reduction is also one of our highest priorities.

In energy engineering and building management, energy efficiency is very often considered the highest priority. But consider this. If you are losing a lot of energy in the building and you install a new high-efficiency HVAC system, then good for you, you are improving the bottom line and using less electricity in the process. But if your envelope is low-performing, leaking air and energy, then you will be paying less money to lose the same amount of air and energy through the envelope that you were before. You are still paying for the mitigation of wasted energy. And spending less money to waste a resource is not being less wasteful of the resource.

Waste reduction has an interesting effect on *upstream* and *downstream* portions of the entire process. *Upstream* refers to the resource flows of energy, water and materials required to construct, restore, retrofit or renovate the building. *Downstream* refers to the resource flows required to operate & maintain the building (and also, arguably, to dispose of it). *Planning for reducing resource flows upstream automatically reduces resource flows downstream.*

Figure 11-2 gives a way to look at it, a cascading effect, which also includes the very important factor of *cost*.

Concerning the long-term operation of buildings, when we continually cut waste upstream, it has a continuous downstream effect

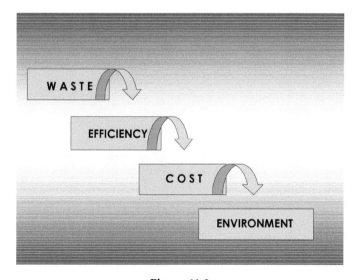

Figure 11-2

which increases efficiency, improves costs and automatically reduces environmental impact over the entire life-cycle of the building.

So, refining our waste reduction and energy efficiency strategies locally with specific building types can prepare us for broadcasting and deploying the same strategies regionally or nationally. This can then become part of a national commitment to efficient, cost-effective and long-term resource management.

We know that the transformation of the energy infrastructure into a clean, secure, efficient and diversified platform for growth is a necessity, and it must be a national priority. So we must prepare ourselves for this transition, and therefore making our buildings high-performance is not just a challenge. It's not just a goal. It's about collaboration, team building, and setting the best possible example for future generations.

Reference

1. http://smartvortex.eu/?q=Project

Sources

GENERAL

Architectural Graphic Standards, 7th Edition, John Wiley, NY, 1981.

ASHRAE: Handbook of Fundamentals, 2007, Atlanta, GA.

Building Science Corp.
http://www.buildingscience.com/

Center for Sustainable Building Research
http://www.csbr.umn.edu/

Cradle to Cradle, McDonough & Braungart, North Point Press, NY, 2002.

Distributed Energy: The Journal of Energy Efficiency & Reliability
Available from: http://www.distributedenergy.com/DE/DEhome.aspx

ENERGY - International, Multi-Disciplinary Journal in Energy Engineering and Research, Editor: H. Lund
http://www.journals.elsevier.com/energy/

Energy Engineering: Journal of the Association of Energy Engineers
Available from: http://www.aeecenter.org/i4a/pages/index.cfm?pageID=3372

Energy in Nature and Society, Smil, MIT Press, MA, 2008.

Energy, Dukert, Greenwood Press, CT, 2009.

Fundamentals of Engineering Materials, Thornton & Colangelo, Prentice Hall, NJ, 1985.

Green Facilities Handbook: Simple & Profitable Strategies for Managers, Woodruff, E., CRC/Fairmont, GA, 2009.

High-Performance Building, Lerum, John Wiley, NJ, 2008.

How to Finance Energy Management Projects, Woodruff, E., Thumann, A., CRC/Fairmont Press, GA, 2013.

Introduction to Architectural Science, Szokolay, Elsevier-Architectural Press, MA,
2008.

IOP Science: On-Line Journal Platform
http://iopscience.iop.org/

Leading Change Toward Sustainability, Doppelt, Greenleaf, Sheffield, UK, 2010.

MarketsandMarkets – Market Research Reports/Renewable Energy Technology
http://www.marketsandmarkets.com/

Matter and Energy, Soddy, Henry Holt, NY, 1912.

Mechanical and Electrical Equipment for Buildings, 11ᵗʰ Edition, Grondzik, Kwok, Stein & Reynolds, John Wiley, NJ, 2010.

Recommended Practice of Daylighting, IES Daylighting Committee, Illuminating Engineering Society of North America, NY, 1979.

Reinventing Fire – Bold Business Solutions for the New Energy Era, Lovins – Rocky Mountain Institute, Chelsea Green, VT, 2011.

Science Direct – Full Text Scientific Database
http://www.sciencedirect.com/

Structure Magazine:
http://www.structuremag.org/index.aspx

The Idea Factory: Bell Labs and the Great Age of American Information, Gertner, Penguin, NY, 2012.

The Prize, Yergin, Free Press, NY, 1992.

The Recycler's Manual for Business, Government and the Environmental Community, Powelson & Powelson, Van Nostrand Reinhold, NY, 1992.

Chapter 1—INPUTS

Annual Energy Review 2011:
http://www.eia.gov/totalenergy/data/annual/pdf/aer.pdf

Buildings Energy Data Book
http://buildingsdatabook.eren.doe.gov/

Department of Energy
http://energy.gov/

Lawrence Livermore National Laboratory
https://www.llnl.gov/

U.S. Energy Information Administration
http://www.eia.gov/

United Nations Human Development Report
http://hdr.undp.org/en/

US Geological Survey - Water Usage
http://water.usgs.gov/watuse/

World Bank World Development Indicators
http://data.worldbank.org/data-catalog/world-development-indicators

WTRG Economics
www.wtrg.com

Chapter 2—DEFINITIONS

High Performance Buildings Magazine – ASHRAE
http://www.hpbmagazine.org/

Whole Building Design Guide
http://www.wbdg.org/

Chapter 3—HIGH-PERFORMANCE: A REAL ESTATE PERSPECTIVE

Building Owners and Managers Association International (BOMA)
http://www.boma.org/Pages/default.aspx

Doing Well by Doing Good? Green Office Buildings, Eichholtz, Piet, Nils
Kok, and John M. Quigley. 2010. *American Economic Review,* 100(5): 2492-
2509.
http://www.aeaweb.org/articles.php?doi=10.1257/aer.100.5.2492

2010 Quadrennial Defense Review, United States Department of Defense,
84-88.
http://www.defense.gov/qdr/images/QDR_as_of_12Feb10_1000.pdf

2011 National Military Strategy of the United States of America, The Joint
Staff, 2.
http://www.jcs.mil/content/files/2011-02/020811084800_2011_NMS_-
_08_FEB_2011.pdf

Energy for the Warfighter: Operational Energy Strategy, United States
Department of Defense, May 2011.
http://energy.defense.gov/OES_report_to_congress.pdf

United States Green Building Council
http://www.usgbc.org/

US Social Investment Network – CERES
http://www.ceres.org/

Value Beyond Cost Savings: How to Underwrite Sustainable Properties,
Green Building Finance Consortium, Scott Muldavin, 2010. *see:* www.
greenbuildingfc.com

The Journal of Sustainable Real Estate
http://www.josre.org/

World Green Building Council
http://www.worldgbc.org/

Chapter 4—BUILDING ENERGY ANALYSIS

ATHENA Life-Cycle Asessment Software: Athena Sustainable Materials Institute
http://www.athenasmi.org/our-software-data/overview/

Building Energy Software Tools Directory
http://apps1.eere.energy.gov/buildings/tools_directory/subjects.
cfm/pagename=subjects/pagename_menu=whole_building_analysis/
pagename_submenu=energy_simulation

DOE-2
http://doe2.com/

Energy Audit of Building Systems: An Engineering Approach. Krarti, M.,
CRC Press, 2000.

Energy Plus
http://apps1.eere.energy.gov/buildings/energyplus/

eQUEST
http://www.doe2.com/equest/

Federal Renewable Energy Screening Assistant (FRESA)
https://www3.eere.energy.gov/femp/fresa/

Life Cycle Analysis – Products & Systems, Pré SIMAPRO
http://www.pre-sustainability.com/simapro-lca-software

Life-Cycle in Sustainable Architecture (LISA)
http://www.lisa.au.com/

National Fenestration Rating Council
http://www.nfrc.org/

Procedures for Commercial Building Energy Audits, 2nd Edition, American Society of Heating Refrigerating and Air-Conditioning Engineers, 2011.

Rocky Mountain Institute: Retrofit Tools
http://www.rmi.org/retrofit_depot_tools_and_resources

Washington State University Cooperative Extension Energy Program
http://www.energy.wsu.edu/

Whole Building Energy Analysis Software Tools – Retrofit
http://apps1.eere.energy.gov/buildings/tools_directory/subjects.cfm/pagename=subjects/pagename_menu=whole_building_analysis/pagename_submenu=retrofit_analysis

Chapter 5—COMMISSIONING

Building Commissioning: WBDG Project Management Committee, June, 2012.
http://www.wbdg.org/project/buildingcomm.php

Building Commissioning Guide: General Services Administration, April, 2005
http://www.wbdg.org/ccb/GSAMAN/buildingcommissioningguide.pdf

Commercial Building Energy Consumption Survey (CBECS)
http://www.eia.gov/consumption/commercial/

California Commissioning Guide: New Buildings, Report Prepared for: The California Commissioning Collaborative, Haasl, T., Heinemeier, K., Portland Energy Conservation, Inc., 2006.
http://www.documents.dgs.ca.gov/green/commissionguideexisting.pdf

Building Commissioning - A Golden Opportunity for Reducing Energy Costs and Greenhouse Gas Emissions, Report Prepared for: California Energy Commission Public Interest Energy Research, Mills, E., Lawrence Berkeley National Laboratory, 2009.
http://cx.lbl.gov/documents/2009-assessment/lbnl-cx-cost-benefit.pdf

Performance Tracking: A Key Part of Ongoing Commissioning, Presentation for Energy Star Monthly Partner Web Conference, Moser, D., Portland Energy Conservation, Inc., August, 2010.
http://www.energystar.gov/ia/business/networking/presentations_2010/Aug10_On-Going_Commissioning.pdf

Principles of Building Commissioning, Grondzik, W., John Wiley & Sons, 2009.

A Retro-commissioning Guide for Building Owners, Portland Energy Conservation, Inc. (with funding from the US EPA ENERGY STAR® Program), 2007.
http://www.peci.org/sites/default/files/epaguide_0.pdf

Texas A&M Energy Systems Laboratory Study. 2009.
http://esl.tamu.edu/

Chapter 6—STANDARDS CODES & RATINGS

American National Standards Institute
http://www.ansi.org/

American Society of Heating, Refrigeration & Air Conditioning Engineers
https://www.ashrae.org/

American Society of Testing Materials
http://www.astm.org/

ASHRAE Standard 189.1
https://www.ashrae.org/resources--publications/bookstore/standard-189-1

ASHRAE Standard 90.1
https://www.ashrae.org/resources--publications/bookstore/standard-90-1

Better Buildings Challenge
http://www4.eere.energy.gov/challenge/

Energy Star
http://www.energystar.gov/

Energy Star Portfolio Manager
http://www.energystar.gov/index.cfm?c=evaluate_performance.bus_
portfoliomanager

Green Building Alliance- DASH (Database for Analyzing Sustainable &
High-Performance Buildings)
http://www.gbapgh.org/content.aspx?ContentID=92

Green Building Initiative
http://www.thegbi.org/

International Energy Conservation Code
http://shop.iccsafe.org/2012-international-energy-conservation-code-
soft-cover.html

International Green Construction Code
Ihttp://www.iccsafe.org/cs/IGCC/Pages/default.aspx

International Standards Organization
http://www.iso.org/iso/home.html

National Institute of Building Sciences
http://www.nibs.org/

National Institute of Standards & Technology
http://www.nist.gov/index.html

United States Green Building Council
http://www.usgbc.org/

Chapter 7—ENVELOPE

American Institute of Architects – Top Ten Metrics
http://www.aiatopten.org/

ArchiExpo – Virtual Architecture Exposition
http://www.archiexpo.com/

Autodesk/Ecotect
http://usa.autodesk.com/ecotect-analysis/

AWS Truepower – Wind Resource Maps
http://www.awstruepower.com/

Buildings Energy Databook – Energy Intensity of the Buildings Sector
http://buildingsdatabook.eren.doe.gov/ChapterIntro1.aspx

Commercial Building Toplighting: Energy Saving Potential and Potential Paths Forward
http://apps1.eere.energy.gov/buildings/publications/pdfs/
commercial_initiative/toplighting_final_report.pdf

Cool Roof Calculator - Department of Energy
http://www.ornl.gov/sci/roofs+walls/facts/CoolCalcEnergy.htm

Cool Roof Rating Council
http://www.coolroofs.org/

Cradle to Cradle Certification
http://www.c2ccertified.org/

Daylighting Design Guide: The Integrated Approach , Lawrence
Berkeley National Laboratory
http://windows.lbl.gov/daylighting/designguide/dlg.pdf

Department of Energy – Energy Analysis
http://www1.eere.energy.gov/analysis/eii_index.html

Department of Energy – Energy Efficiency & Renewable Energy/
Building Science Explorer
http://basc.pnnl.gov/building-science-explorer

eBUILD – Online New Building Products
http://www.ebuild.com/

Efficient Windows Collaborative
http://www.efficientwindows.org/

Foundation Design Handbook – Oak Ridge National Laboratory
http://www.ornl.gov/sci/buildingsfoundations/handbook/index.
shtml

Green Roof Systems: Standard Guide for Selection, Installation, and
Maintenance of Plants for : ASTM E2400 – 06
http://www.astm.org/Standards/E2400.htm

Green (Vegetative) Roof Systems: Standard Practice for Determination of Dead
Loads and Live Loads Associated with, ASTM E2397 – 11
http://www.astm.org/Standards/E2397.htm

GreenPoint Manufacturing and Design Center
http://www.gmdconline.org/

Institute for Research in Construction, National Research Council of
Canada
http://archive.nrc-cnrc.gc.ca/eng/ibp/irc.html

Insulating Concrete Form Association
http://www.forms.org/

International Green Roofing Association (IGRA) – Publications
http://www.igra-world.com/green_roof_literature/index.php

Life-Cycle Costs of Buildings and Building Systems - Standard Practice for
Measuring: ASTM E917 - 05(2010)
http://www.astm.org/Standards/E917.htm

National Solar Radiation Database
http://rredc.nrel.gov/solar/old_data/nsrdb/1991-2010/

National Weather Service/NOAA – Rainfall Maps
http://www.weather.gov/

Polyisocyanurate Manufacturers' Association (PIMA) – Technical
Bulletins
http://www.polyiso.org/?page=TechnicalBulletins

Radiance Lighting Simulation Program – LBNL
http://radsite.lbl.gov/radiance/HOME.html

Reflective Roof Coatings Institute
http://www.therrci.org/

Roof Coating Manufacturers' Association
http://roofcoatings.org/

Roof Savings Calculator (RSC) – Lawrence Berkeley Laboratory, Oak
Ridge National Laboratory
http://www.roofcalc.com/

Structural Insulated Panel Association
http://www.sips.org/

University of California at Los Angeles – Energy Design Tools
http://www.energy-design-tools.aud.ucla.edu/

US Library of Congress – "Built in America" Collection
http://www.loc.gov/pictures/collection/hh/

Whole Building Design Guide
http://www.wbdg.org/

Windrose PRO – Enviroware
http://www.enviroware.com/portfolio/windrose-
pro/#axzz1gCPM7IqK

Wolfram Research – Physics/Energy
http://scienceworld.wolfram.com/physics/topics/Energy.html

Zero Energy Buildings Database
http://zeb.buildinggreen.com/

Sound: References

Acoustical Privacy in the Landscaped Office Warnock, A.C.C., J. Acoust. Soc.
Am., 53(6), pp. 1535-1543 (1973).

Annual Book of ASTM Standards, Vol. 04.06, pp. 780-787
ASTM E492, Standard Test Method for Laboratory Measurement of Impact Sound Transmission through Floor- Ceiling Assemblies using the Tapping Machine. 1997

ASTM E989, Standard Classification for Determination of Impact Insulation Class (IIC). 1997 Annual Book of ASTM Standards, Vol. 04.06, pp. 851-853.

Controlling the transmission of airborne sound through floors, Warnock, A.C.C., Institute for Research in Construction, National Research Council of Canada, Construction Technology Update 25, 1999

Designing Quiet Structures: A Sound Power Minimization Approach, Koopmann, G., Director, Center for Acoustics and Vibration, The Pennsylvania State University, Fahnline, J., Director, Center for Acoustics and Vibration, The Pennsylvania State University, Academic Press, 1997. http://www.elsevier.com/books/designing-quiet-structures/koopmann/978-0-12-419245-4#

Effect of electrical outlet boxes on sound insulation of a double leaf wall, Nightingale, T.R.T. IRC-IR-772, National Research Council, October 1998

Factors affecting sound transmission loss., Warnock, A.C.C. Institute for Research in Construction, National Research Council, Canadian Building Digest 239, Ottawa, 1985

Guideline on Office Ergonomics, Warnock, A.C.C. and Birta, J.A. CSA Standard Z412-00 (2000).

Sound isolation and fire resistance of assemblies with firestops. Nightingale, T.R.T. and Sultan, M.A. Construction Technology Update No. 16, Institute for Research in Construction, National Research Council of Canada, 1998.

Sound transmission loss of masonry walls: twelve-inch lightweight concrete blocks — comparison of latex and plaster sealers., Northwood, T.D. , Monk, D.W. Division of Building Research, National Research Council,

Building Research Note 93. September 1974

Summary Report for Consortium on Fire Resistance and Sound Insulation of Floors: Sound Transmission Class and Impact Insulation Class Results. Institute for Research in Construction, National Research Council of Canada, Internal Report 766, 121 p., April 1999.

The Acoustical Design of Conventional Open Plan Offices, Bradley, J.S. Canadian Acoustics, 27 (3) 23-30 (2003).

Chapter 8—ON-SITE POWER (DEG)

American Wind Energy Association
http://www.awea.org/

Assessment of Distributed Generation Technology Applications, Resource Dynamics Corporation , 2001. See: http://www.distributed-generation. com/Library/Maine.pdf

Distributed Generation: a Definition, Ackermann, T., Andersson, G., Söder, L. Electric Power Systems Research, Vol. 57, 2001, pp. 195-204.

Distributed Energy Generation and Sustainable Development, Kari Alanne, Arto Saari, Laboratory of Construction Economics and Management, Department of Civil and Environmental Engineering, Helsinki University of Technology, P.O. Box 2100, 02015 HUT, Finland.
http://www.sciencedirect.com/science/article/pii/S1364032105000043

Distributed Generation: The Power Paradigm for the New Millennium, Borbely, A., Alexandria, Virginia, USA; Kreider, J. F., Boulder, CO, USA, CRC Press, 2001.
http://www.crcpress.com/product/isbn/9780849300745

DOE Fuel Cell Technologies Program Record http://www.hydrogen.energy. gov/pdfs/12020_fuel_cell_system_cost_2012.pdf

Electric Power Research Institute: http://www.epri.com/search/Pages/ results.aspx?k=recuperated%20microturbine%20system

EPA – Combined Heat & Power Partnership – Catalogue of CHP Technologies
http://www.epa.gov/chp/technologies.html

Fuel Cells – Fundamentals and Applications, Carette, Friedrich, Stimming, WILEY-VCH Verlag GmbH, Weinheim, Fed. Rep. of Germany, 2001.
http://onlinelibrary.wiley.com/doi/10.1002/1615-6854(200105)1:1%3C5::AID-FUCE5%3E3.0.CO;2-G/full

Integration of Distributed Generation in the Power System (IEEE Press Series on Power Engineering), Bollen, H., Hassan, F., Wiley-IEEE Press, August, 2011.

Massachusetts Institute of Technology – Cogeneration Project
http://cogen.mit.edu/index.cfm

National Renewable Energy Laboratory – Photovoltaics Research
http://www.nrel.gov/pv/

Renewable Energy : An International Journal, The Official Journal of WREN - The World Renewable Energy Network, editor: A.A.M. Sayigh
http://www.journals.elsevier.com/renewable-energy/

Renewable Energy: Power for a Sustainable Future, Boyle, G., Oxford University Press, USA,November, 2012.

The Role of Energy Storage with Renewable Electricity Generation, NREL Technical Report, Denholm, P., Ela, E., Kirby, B., Milligan, M., 2010.
http://www.nrel.gov/docs/fy10osti/47187.pdf

Solar Energy: The Official Journal of the International Solar Energy Society®, Editor-in-Chief: Y. Goswami
http://www.journals.elsevier.com/solar-energy/

United States Clean Heat & Power Association
http://www.uschpa.org/i4a/pages/index.cfm?pageid=1

Wind Energy Engineering, Pramod Jainhttp://www.amazon.com/Wind-Energy-Engineering-Pramod-Jain/dp/0071714774/ref=sr_1_13?s=books&ie=UTF8&qid=1368019937&sr=1-13&keywords=small+wind+energy - #, Mcraw-Hill, 2010.
http://accessengineeringlibrary.com/browse/wind-energy-engineering

Wind Research – National Renewable Energy Laboratory
http://www.nrel.gov/wind/publications.html

Chapter 9—MECHANICAL & ELECTRICAL EQUIPMENT

Air-Conditioning, Heating & Refrigeration Institute
http://www.ahrinet.org/

ASHRAE: Handbook of Fundamentals, 2007, Atlanta, GA.

ANSI/ASHRAE/IES Standard 90.1-2010, *Energy Standard for Buildings*,
Atlanta, GA.

ANSI/ASHRAE/USGBC/IES Standard 189.1-2011, *Standard for the
Design of High Performance Buildings*, Atlanta, GA.

ASHRAE Standard 62.1 -2010: *Ventilation for Acceptable Indoor Air
Quality*, Atlanta, GA.

Better Buildings Alliance
http://www4.eere.energy.gov/alliance/node/9

Buildings Energy DataBook – Buildings Sector Energy Consumption
http://buildingsdatabook.eren.doe.gov/ChapterIntro1.aspx?1%20
-%201

*Consumer's Directory of Certified Efficiency Ratings for Heating and Water
Heating Equipment.* AHRI
http://www.ahridirectory.org/ahridirectory/pages/home.aspx

Commercial Energy Auditing Reference Handbook, 2nd Edition, Doty, S.,
Fairmont Press, GA, 2011.

Efficient Lighting Applications & Case Studies, Dunning, S., Thumann, A.,
Fairmont Press, GA, 2013.

Electric Lighting Controls, Whole Building Design Guide, *http://www.
wbdg.org/resources/electriclighting.php*

Energy Management Handbook, 8th Edition, Doty, S., Turner, W., Fairmont Press, GA, 2012.

Energy.Gov – Furnaces & Boilers
http://energy.gov/articles/furnaces-and-boilers

Guide to Energy Management, 7th Edition, Capehart, B. ,Turner, W., Kennedy, W., Fairmont Press, GA, 2011.

Illuminating Engineering Society
http://www.iesna.org/

Introduction to Premium Efficiency Motors, Copper Development Association, 2013, http://www.copper.org/environment/sustainable-energy/electric-motors/education/motor_text.html

Mechanical and Electrical Equipment for Buildings, 11th Edition, Grondzik, W., Kwok, A., Stein, B., Reynolds, J., John Wiley, NJ, 2010.

A Meta-Analysis of Energy Savings from Lighting Controls in Commercial Buildings , Williams, A., Atkinson, B., Garbesi, K., Rubinstein , F., Energy Analysis Department, Lawrence Berkeley National Laboratory, September 2011, http://efficiency.lbl.gov/drupal.files/ees/Lighting%20Controls%20in%20Commercial%20Buildings_LBNL-5095-E.pdf

Practical Controls: A Guide to Mechanical Systems, Calabrese, S., CRC/Fairmont Press, Atlanta, GA, 2003.

Understanding & Designing Outdoor Air Systems (DOAS), Mumma, S., ASHRAE Short Course, 2010, http://www.caee.utexas.edu/prof/Novoselac/classes/ARE389H/Handouts/doas_arkansas_final.pdf

Underfloor Air Technology
http://www.cbe.berkeley.edu/underfloorair/Introduction.htm

Chapter 10—OPERATIONS & MAINTENANCE

American Membrane Technology Association
http://www.amtaorg.com/

Arizona Municipal Water Users Association
http://www.amwua.org/

Container Recycling Institute
http://www.container-recycling.org/

Energy Informatics
http://energyinformatics.info/

Facilities.net
http://www.facilitiesnet.com/bom/default.asp

Federal Energy Management Program – Operations & Maintenance Best Practices
http://www1.eere.energy.gov/femp/pdfs/omguide_complete.pdf

Federal Leadership in Environmental, Energy & Economic Performance
http://www.whitehouse.gov/assets/documents/2009fedleader_eo_rel.pdfGlobal

Environmental Management Initiative
http://www.gemi.org/gemihome.aspx

Leadership in Energy & Environmental Design– Existing Buildings, Operations & Maintenance (LEED-EB O&M)
http://www.usgbc.org/sites/default/files/LEED%202009%20Rating_EBOM-GLOBAL_07-2012_8d_0.pdf

National Sanitation Foundation
http://www.nsf.org/

Sustainable Facilities Tool – Life-Cycle Assessment
http://sftool.gov/plan/400/life-cycle-assessment-lca-overview

Index